SELECTED WRITINGS ON THE

HISTORY OF SCIENCE

Classics of British Historical Literature

JOHN CLIVE, EDITOR

William Whewell

——

Selected Writings on the History of Science

EDITED AND WITH AN INTRODUCTION BY

YEHUDA ELKANA

The University of Chicago Press

CHICAGO AND LONDON

The engravings and woodcuts in
this edition did not appear
in the original selections.

The University of Chicago Press, Chicago 60637
The University of Chicago Press, Ltd., London
© 1984 by The University of Chicago
All rights reserved. Published 1984
Printed in the United States of America
91 90 89 88 87 86 85 84   1 2 3 4 5

Library of Congress Cataloging in Publication Data

Whewell, William, 1794–1866.
  Selected writings on the history of science.

  (Classics of British historical literature)
  Bibliography: p.
  1. Science—History.   I. Elkana, Yehuda, 1934–
II. Title.   III. Series.
Q125.W56   1984   509   83-9187
ISBN 0-226-89433-9
ISBN 0-226-89434-7 (pbk.)

There is a mask of theory over the whole face of Nature

William Whewell
Philosophy of the Inductive Sciences

Contents

Series Editor's Preface

This series of reprints has one major purpose: to put into the hands of students and other interested readers outstanding—and sometimes neglected—works dealing with British history which have either gone out of print or are obtainable only at a forbiddingly high price.

The phrase Classics of British Historical Literature requires some explanation, in view of the fact that the two companion series published by the University of Chicago Press are entitled Classic European Historians and Classic American Historians. Why, then, introduce the word *literature* into the title of this series?

One reason is obvious. History, if it is to live beyond its own generation, must be memorably written. The greatest British historians—Clarendon, Gibbon, Hume, Carlyle, Macaulay—survive today, not merely because they contributed to the cumulative historical knowledge about their subjects, but because they were masters of style and literary artists as well. And even historians of the second rank, if they deserve to survive, are able to do so only because they can still be read with pleasure. To emphasize this truth at the present time, when much eminently solid and worthy academic history suffers from being almost totally unreadable, seems worth doing.

The other reason for including the word *literature* in the title of the series has to do with its scope. To read history is to learn about the past. But if, in trying to learn about the British past, one were to restrict oneself to the reading of formal works of history, one would miss a great deal. Often a historical novel, a sociological inquiry, or an account of events and institutions couched in semifictional form teaches us just as much about the past as does the "history" that calls itself by that name. And, not infrequently, these "informal" historical works turn out to be less well known than their merit deserves. By calling this series Classics of British Historical Literature it will be possible to include such books without doing violence to the usual nomenclature.

The history of science, now a flourishing field, is a very recent newcomer to historiography. It would not be unjust to call it a twentieth-century branch of historical scholarship. Nevertheless, it has its pioneers. And there is no doubt that one of the chief among them was William Whewell. His interest in the subject certainly reflected that optimistic sense of progress which is so often, and rightly, held to be a hallmark of the nineteenth century in Britain. In the introduction to his *History of the Inductive Sciences*, written in the year of Queen Victoria's accession, Whewell proudly points to the fact that his own generation was heir to a vast patrimony of science. Our species, he writes, ever since the time of its creation has been traveling onward in pursuit of truth, "and now that we have reached a lofty and commanding position, with the broad light of day around us, it must be grateful to look back on the line of our past progress."

If that statement seems to sound all too crassly the clarion call of the Whig interpretation of history, Professor Elkana amply demonstrates in his introductory essay to this volume that Whewell was very far from being a naïve celebrant of unremitting scientific progress. He was, in fact, very much aware of the importance of failures and controversies for the historian of science. In that respect, as well as in his anticipations of Popper's critique of logical positivism and in his emphasis on the scientific innovator's need to combine thought with observation and experiment, much of what he wrote has a very modern sound.

The selections in the present volume succeed in giving the reader a good sense of Whewell's scope and depth as a historian and philosopher of science. They also convey some idea of his strength as a writer of clear and beautiful prose. Take this passage on the genius of Newton:

> The hidden fountain of our unbidden thoughts is for us a mystery; and we have, in our consciousness, no standard by which we can measure our own talents; but our acts and habits are something of which we are conscious; and we can understand, therefore, how it was that Newton could not admit there was any difference between himself and other men, except in his possession of such habits as we have mentioned, perseverance and vigilance.

It is given to few historians to be both profound and readable. Whewell was one of those few.

JOHN CLIVE

Editor's Introduction

The strain of music from the lyre of Science flows on rich and sweet,
full and harmonious, but never reaches a close: no cadence is heard
with which the intellectual ear can feel satisfied.

William Whewell,
Indications of the Creator

William Whewell was born on 24 May 1794 in Lancaster. His
father was a well-to-do carpenter. After attending the Blue School,
he went to the Lancaster Grammar School, where the later celebrated naturalist, Richard Owen, was his schoolfellow. In 1810 he
transferred to Haversham Grammar School in order to compete for
a fifty-pound exhibition to Trinity College, Cambridge. He was
successful.

At Cambridge, where he arrived in 1811, he read broadly in
many areas and even won first prize for a poem. But, already, his
interests were beginning to focus on natural philosophy. In a letter
to his father written during his third year at the university (1814),
Whewell relates that the students had received their prize book
money, which amounted to twenty shillings each, enough "to buy
a book, get it bound, gilt and lettered with the college arms."
Instead, he himself had bought for ten pounds Newton's *Principia*,
"a book that I should unavoidably have to get sooner or later," in
three quarto volumes most elegantly bound. He occasionally went
to London to go to the theater. He read German philosophy. He
got involved in debates about the Lake poets and took Coleridge's
side against Wordsworth, whom he learned to appreciate only
many years later. In 1817 he became president of the Union at
Cambridge, at that time still considered an illegal gathering. He
was presiding when the vice-chancellor and proctors came into the
building in order to break up one of the meetings. Even then, his
authority was considerable. He demanded that the proctors with-
draw so that the society could consult on its response.

In 1819, Whewell was among the founders of the Cambridge
Philosophical Society. A year later he was elected a fellow of the
Royal Society, after publishing the important *Elementary Treatise on*

Mechanics (1819)—a treatise which, following on the missionary work of Whewell himself and of Herschel, Peacock, and Babbage at Cambridge to modernize the teaching of mathematics, introduced the Continental calculus into physics, replacing the system of fluxions. The book was acclaimed for its clearness, correctness, and advanced mathematics. Whewell spent the succeeding years writing a number of textbooks in mechanics and dynamics, some works on mineralogy, and—this shows the breadth of his interests—a volume entitled *Architectural Notes on German Churches* (1830). In 1826 he was ordained as a priest in the Church of England. The following year he was elected to the Geological Society, and in 1828, after three years of legal difficulties about the mode of election to the Chair, he was elected Professor of Mineralogy.

Whewell and his friend, G. B. Airy, had traveled to Cornwall in 1826 in order to take some gravitation measurements in Dolcoath Mine, with the purpose of determining the density of the earth. It was characteristic of Whewell to have exploited the more dramatic aspects of that experiment. On 10 June 1826, he wrote a letter to his friend Lady Malcolm datelined "Underground Chamber, Dolcoath Mine, Camborne, Cornwall":

> I venture to suppose that you never had a correspondent who at the time of writing was situated as your present one is. I am at the moment sitting in a small cavern deep in the recesses of the earth, separated by 1200 feet of rock from the surface on which you mortals tread.

A vivid, detailed description of the people at the bottom of the mine and the machinery being used followed.

Whewell published his Bridgewater Treatise, *Astronomy and General Physics Considered with Reference to Natural Theology* in 1833. In it, he showed that great scientific discoverers all believed in an intelligent Maker of the universe, while the doubters contributed little to science. This was the first work that brought Whewell wide recognition. It lies at the crossroads of two lines of thought which preoccupied him: the deeply religious wish to prove the argument for the existence of God from the orderly design of the universe, and his historical approach to events. Even in this treatise, which is primarily theological, the argument is fundamentally historical. It relates the stories of discoverers and their discoveries in order to draw religious lessons from history.

In 1836, at the urgent request of his friends Sedgwick, Lyell, and Murchison, Whewell accepted the presidency of the Geological Society. A year later he received the Royal Prize Medal from the Royal Society for his work on tides. In 1838, he resigned the Chair of Mineralogy in order to assume the vacant Chair of Moral Theology. Characteristically, Whewell preferred the title Professor of Moral Philosophy. The first edition of his magnum opus, the *History of the Inductive Sciences*, appeared in three volumes in 1837. It was widely reviewed and acclaimed, though it also received some critical comment. In quick succession to the *History*, the two-volume *Philosophy of the Inductive Sciences* appeared in 1840. While he was in the process of publishing these major works, Whewell also published, as a *jeu d'esprit*, a translation of Goethe's *Hermann and Dorothea* (1839).

The year 1841 was a time for rethinking his life and work. He felt that he had given all he had to give as a resident fellow at Trinity and wrote to a friend that "college rooms were no home for declining years." (He was forty-seven!) He explored the possibilities of taking up a college living, and visited Marsham with this in mind, only to reach the conclusion that it was not for him. To his friend Archdeacon Hare, he wrote, "On the other hand, am I fit to take a cure of souls? I like to have the disposal of my time."

Then his personal situation changed. He met and soon decided to marry Cordelia Marshall of Leeds. It turned out to be a very happy marriage. On the day of the wedding, 12 October 1841, Whewell received a letter from the master of Trinity, Christopher Wordsworth, announcing his resignation and expressing the wish that Whewell would become his successor. That wish, of course, was not sufficient to obtain the position for him. The appointment was in the gift of the queen. But even before Whewell had indicated his own interest in the position, Sir Robert Peel, the prime minister, had recommended him to Queen Victoria for the mastership. He described Whewell in these terms:

> A general favorite among all who have intercourse with him for his good temper and easy and conciliatory manners. Though not particularly eminent as a divine (less so at least than as a writer on scientific and philosophical subjects) his works manifest a deep sense of the importance of religion and sound religious views.

Whewell accepted the position; and he and his wife began a new life in Trinity Lodge.

Whewell not only presided over the college, he also became

deeply involved in Cambridge University affairs. He was twice vice-chancellor, once in 1842 and again in 1856. He continued to write major philosophical and scientific papers, though no full-scale books, during these years. Not surprisingly, even his enormous energies became overtaxed. We know that his close friend Hare reprimanded him at one point for his high-handedness and his natural vehemence of character, about which Hare had heard from various quarters. Whewell accepted the rebuke in a friendly spirit but denied its substance. (One factor which may have contributed to his making some enemies was his inability to remember faces.) Nonetheless, the years after his assumption of the mastership witnessed intensive intellectual activity as well as a happy and relaxed family life. With Whewell at its head, Trinity College once again became the center of academic and political influence in Cambridge.

A number of Whewell's publications were accompanied by intense controversies, some of which became famous. In 1853, he published anonymously a book on the plurality of worlds which led to a long battle with David Brewster and others. He also became involved in the question of university reform, as it was argued in the fifties, as well as in a controversy with John Stuart Mill over induction.

In 1855, convinced that he had said all he could on the subject of moral philosophy, he resigned his chair. In the same year, his beloved Cordelia died, and for some months thereafter he was on the verge of clinical depression. His large circle of friends helped to sustain him during this period. Indeed, the list of those whom Whewell used to call "friends of a lifetime" was both long and formidable. It included Herschel, Jones, Sedgwick, Worsley, Rose, Peacock, Hare, Kenelm Digby, Airy, Henslow, James Forbes, as well as many others. His warmth and his need for both receiving and giving love were remarkable; and although he participated in many a lively controversy, none of his friendships broke under the strain, and there were no men or women, young or old, who he cultivated and later dropped or forgot. His personality was very powerful. People either liked or disliked him. No one was neutral toward him.

Whewell always took a great interest in legal philosophy and in comparative law. He dealt with legal issues in his *Elements of Morality, Including Polity* (1845), edited the works of the great seventeenth-century lawyer Hugo Grotius, and in 1847, when the Chair of Civil Law at Cambridge became vacant, was actively involved in an appointment which would elevate civil law into the general study of jurisprudence. The nomination went to the eminent, anthropologically minded student of comparative law, Sir Henry Maine. (In his will, Whewell made provision for the study of international law at Cambridge by establishing the Whewell Chair.)

Whewell's unusual breadth of knowledge may be appreciated when one reads his exchange of letters with James Spedding on the latter's edition of the works of Bacon. Spedding had written a two-volume refutation of Macaulay's attack on Bacon's character. Whewell read the work, and found that Spedding claimed not to have understood Bacon's remark to the effect that "of all the nations the English are not subject, base nor taxable." He supplied the information that Bacon's reference must have been "to the description given by the French people in their law books: that they are "peuple bas, soumis (or some such word), *taillable, et corvéable à merci et à miséricorde.*"

In 1858, Whewell married Lady Affleck, the widowed sister of Leslie Ellis, Spedding's collaborator on the edition of Bacon. Once again, this was a very happy marriage, and one which provided Whewell with a few more years of undisturbed creativity. However, his second wife died in 1865 and from then on Whewell's body and spirit weakened rapidly. On Saturday, 24 February 1866, he fell from his horse and suffered serious injuries, to which he succumbed some days later. He was buried in the antechapel of Trinity College, where he had, in 1842, erected a magnificent marble copy of the statue of Bacon at St. Albans. Whewell's last publication was an article on Grote's Plato which appeared posthumously. Respectful notices of his death were published in all British and in many Continental periodicals. One of the most remarkable was by David Brewster, president of the Royal Society of Edinburgh, who spoke "in gentle and respectful language over the tomb of one with whom he had been so often in controversy." With Whewell's death, all agreed, Victorian England had lost one of its most brilliantly controversial, yet at the same time one of its most typical spokesmen.

Whewell was a polymath—mathematician, architect, physicist, crystallographer, theologian, political economist, and educator. He was primarily a historian, a philosophically minded historian, to be sure, but his emphasis was always on processes, on time-dependent factors, on ever-present change. As he says at the end of his introduction to the *History of the Inductive Sciences:*

> I may further add that the other work to which I refer, the Philosophy of the Inductive Sciences, is in a great measure historical, no less than the present *History*. That work contains the history of the Sciences as far as it depends on *Ideas*; the present work contains the history as far as it depends upon *Observation*.

The rationale of Whewell's reliance on history is that we can learn from it:

> We may best hope to understand the nature and conditions of real knowledge, by studying the nature and conditions of the most certain and stable positions of knowledge which we already possess: and we are most likely to learn the best methods of discovering truth by examining how truths, now universally recognized, have *really* been discovered. [Editor's italics].

Thus the source for philosophical analysis is history, *real* history, and not rational reconstruction. Such history, Whewell wrote, can perhaps give "the best methods for the discovery of new truths," and even "nothing less than a complete insight into the essence and conditions of all real knowledge." However, let us not misunderstand Whewell: the finest methods can, at best, supply us with necessary conditions for scientific discovery—they are only means toward the end of avoiding mistakes due to shortsightedness and lack of methodological usage. The methods in themselves cannot *ensure* discoveries:

> It can hardly happen that work which treats of Methods of Scientific Discovery shall not seem to fail in the positive results which it offers. For an Art of Discovery is not possible. At each step of the progress of science, are needed invention, sagacity, genius;—elements which no Art can give.

A century later, Karl Popper was to write a philosophical masterpiece, *The Logic of Scientific Discovery*, to prove that there was indeed *no* logic of scientific discovery.

Whewell's realism is thoroughgoing, in history as in daily life. Take, as an instance, this passage addressed to Herschel in 1884 in "Remarks on a Review of *The Philosophy of the Inductive Sciences:*"

In short, space is a *reality*, and not a mere matter of convention or imagination. Now, if by calling space an idea, we suggest any doubt of its reality and of the reality of the external world, we certainly run the risk of misleading our readers; for the external world is real if anything be real: the bodies which exist in space are things, if things are anywhere to be found. That bodies do exist in space, and that *that* is the reason why we apprehend them as existing in Space, I readily grant. But I conceive that the term Idea ought not to suggest any such doubt of the reality of the knowledge in which it is involved. Ideas are always, in our knowledge, conjoined with facts. Our real knowledge is knowledge because it involves ideas, real, because it involves facts. We apprehend things as existing in space because they do so exist and our ideas of space enable us to observe them, and so to conceive them.

To which Whewell adds "knowledge requires ideas. Reality requires things."

Although it was conceived and expounded a century before logical positivism, Whewell's thought, astonishingly in advance of its time, strikes us now as a lively refutation of that twentieth-century mode of philosophy. Whewell always emphasized the processes of change, but he believed even more firmly in a permanent evolution of thought and even introduced a dialectical view of the process of the growth of knowledge. His dialectic applies both to Ideas and Facts, and metaphysics plays a central role in its operation. While his basic tenets must, of course, be seen in the context of the early nineteenth century, there can be no doubt about the similarities between his and Popper's much later philosophy of science.

Both the *History* and the *Philosophy* are, as we have already noted, actually histories, albeit philosophical histories. What, then are the Inductive Sciences? They are, in brief, those sciences for which Whewell can claim that "real speculative knowledge demands the combination of the two ingredients: right reason, and facts to reason upon." Sciences which have no facts to reason upon, like the various branches of mathematics, are not included. "Sciences," not "Science." Whewell comments about this as follows:

> We may best hope to make some progress towards the Philosophy of Science by employing ourselves upon the Philosophy of the Sciences.

The history that results from such employment will exhibit the progress of scientific truth. Whewell's organizing idea is to

> throw the history of each science into Epochs at which some great
> and cardinal discovery was made, and to arrange the subordinate
> events of each history as belonging to the Preludes and the Sequels
> of such Epochs.

So far, Whewell is the proud, present-oriented Victorian, always
conscious of progress. However, even here a new tune may be
heard. For, in spite of all his commitment to positive achievement,
Whewell is not purely success oriented like his contemporaries or
his positivistic successors. He knows that there exists a creative
dialectic between success and failure, just as he knows that we learn
from failures no less than from successes. Even that which seems to
us a failure today may, he argues, carry in itself the nucleus of a
truth in the future. He is as interested in studying "instances in
which science has failed to advance" as he is in studying successes.
The study of failures will teach us how "in consequence of the
absence of one or other" of the necessary prerequisites for a discovery, success was prevented. Thus, when he approaches the study of
the Greek school of philosophy, Whewell considers the problem of
why the Greeks failed to develop science. Although they introduced abstract conceptions—"vague, indeed, but not, therefore,
unmeaning"—when they attempted to make these clearer and
more definite, instead of attending to both facts and things, they
confined themselves to examining words:

> The method of real inquiry was the way to success; but the Greeks
> followed the . . . *verbal* or *notional* course, and failed.

In-depth analysis of failures is one of Whewell's lasting achievements as a historian of science. In studying both success and
failure, it is progress, i.e., change, in science that primarily
interests him:

> In our history, it is the progress of knowledge only which we have
> to attend to. This is the *main action* of our drama. [Editor's italics]

The process is dialectical:

> Perhaps one of the most prominent points of this work [*The Philoso-
> phy of the Inductive Sciences*] is the attempt to show the place which
> discussions concerning Ideas have had in the progress of science

. . . The first ten books of the present work . . . contain an account of the principal philosophical controversies which have taken place in all the physical sciences from Mathematics to Physiology . . . and these controversies . . . have been an essential part of the discoveries made.

The dialectical process of critical analysis does not apply to Ideas only, it applies even more to the fundamental antitheses of Philosophy—to Facts and Ideas, or Works and Words, or Things and Thoughts. For, on the one hand, in order to organize our history and our philosophy, we must distinguish between Facts and Ideas; on the other, as we shall soon see, there is no sharp line of demarcation between them. Moreover,

> without Thoughts there could be no connexion; without Things there could be no reality; Thoughts and Things are so intimately combined in our Knowledge, that we do not look upon them as distinct. Our single act of the mind involves them both; *and their contrast disappears in that union*. [Editor's italics]

Few have better described what dialectics is about or have perceived so clearly the need for demarcation:

> But though Knowledge requires the union of these two elements [Facts and Ideas] Philosophy requires the separation of them in order that the nature and structure of Knowledge may be seen.

It is the interaction between Facts and Ideas which constitutes progress. Absence of such interaction, or arrested dialectics, does not mean that either of these elements is absent. It is enough that the act of mind which brings about their unity be absent for any reason whatsoever. Even in the case of the Greeks, when, as we saw above, failure resulted from preoccupation with words, the cause of failure was not neglect of the facts. What, then, went wrong?

> To this I answer: The defect was, that though they had in the possession Facts and Ideas, *the Ideas were not distinct and appropriate to the Facts*.

To coordinate Ideas and Facts in the appropriate dialectic is according to Whewell, the "Colligation of Facts," a concept to which an entire chapter (4) of the second volume of the *Philosophy* is devoted.

A tumultuous ever-active dialectic interaction between Facts and Ideas, successes and failures following upon one another constitute progress in science. In contrast, stationary periods are characterized by an arrest of the dialectical process. In

> stationary periods we shall find that the process we have spoken of as essential to the formation of real science, the conjunction of clear Ideas with distinct Facts, was interrupted, and in such cases, men dealt with Ideas alone.

Ideas must come first, but, if they are not combined with facts, progress is arrested, the period is stationary. When, however, the Colligation of Facts begins, then we get the prelude to an Inductive Epoch that is followed by the Inductive Epoch itself, and afterwards, by the Sequel.

Whewell's philosophy of history is both anthropological and comparative, and it thus constitutes one of the few nineteenth-century English philosophies influenced by Continental thought. Not only is Whewell knowledgeable about the different French, German, Italian, and other schools of philosophy and contributions to science. He has even absorbed the typically Continental form of *Problemstellung:* Thus he asks whether the developments described in his *Progressive History of Science* are universally typical of mankind, or whether they are culture-bound phenomena.

> At an early period of history there appeared in men a propensity to pursue speculative inquiries concerning the various facts and properties of the material world. What they saw excited them to meditate, to conjecture, and to reason.

Yet this universal interest took different forms in different cultures: "The Egyptians, it appears, had no theory, and felt no want of theory. Not so the Greeks."

Whewell rejected the Lockean view, according to which no knowledge could be derived from the mind alone. Ideas, as supplied by the mind itself, must be dialectically combined with Sensation in order to produce Knowledge, making Ideas "Not Objects of Thought but rather Laws of Thought." All knowledge comes through the senses; and no ideas are objects of our thoughts. Locke's empiricist doctrine was still very influential in England in the nineteenth century, and Whewell found it so important to refute him that in the dedication of *The Philosophy of the Inductive Sciences* to his friend and colleague, the geologist the Reverend

Adam Sedgwick, he wrote: "To you I must justly inscribe a work which contains a criticism of the fallacies of the ultra-Lockean school." True to his promise, in writing both the *History* and the *Philosophy*, Whewell never omitted a thorough description of the controversies which accompanied any newly emerging idea or newly conceived Colligation of Facts and Ideas.

The controversies to which Whewell refers as essential parts of discoveries are metaphysical. When Kant built his philosophy around the concept of metaphysics, this was still an acceptable philosophical term and did not symbolize the deep antithesis to science which it came to denote by the time Whewell was writing. For most of the success-oriented, triumphant Victorians, metaphysics was nonsense; or, at best, poetry. Sir John Herschel's *Preliminary Discourse on the Study of Natural Philosophy* (1830) had no place for metaphysics in its view of the world. Whewell, on the other hand, maintains:

> Physical discoverers have differed from barren speculators, not by having no metaphysics in their heads, but by having good metaphysics while their adversaries had bad, and by binding their metaphysics to their physics, instead of keeping the two asunder.

Still, metaphysics was not enough. Progress in science was made through trial and error:

> Advances in knowledge are not commonly made without the previous exercise of some boldness and license in guessing. The discovery of new truths requires undoubtedly, minds careful and scrupulous in examining what is suggested; but it requires, no less, such as are quick and fertile in suggesting. *What is Invention, except the talent of rapidly calling before us many possibilities, and selecting the appropriate one?* [Editor's italics]

Progressive science involves a metaphysical element, Ideas, and a physical element, Facts; and between the two there is the Fundamental Antithesis, dialectically leading to a union.

Whewell's problem of finding a demarcation between physics and metaphysics, between facts and ideas, is made more complex by his awareness that no real distinction can be made. However, the relation is not symmetrical: "the Idea can never be independent of the Fact" (although "the Fact must never be drawn towards the Idea"); but the Fact (in his opinion) may be independent of the

Idea. In the language of contemporary philosophy of science this view may be translated into "all theory is observation-dependent, but observation can be theory-independent." A strange view. Its first part is very historical and has recently been heard from history-minded philosophers, but never in combination with the view that observation (Fact) can be independent of theory, a view which used to be held by the early logical positivists and is rapidly losing supporters. Whewell is aware of this difficulty, too. He asks: "Does not the apprehension of the Fact imply assumptions which may with equal justice be called Theory, and which are perhaps fake Theory?" All ideas, i.e., theory, are observation-laden; just as all fact, i.e., observation, is theory-laden:

> And as perception of Objects implies Ideas,—as Observation implies Reasoning; so, on the other hand, Ideas cannot exist where Sensation has not been; Reasoning cannot go on where there has not been previous Observation.

I do not wish to create the impression that Whewell's ambiguous attitude to the issue of observation-laden theory and theory-laden observation, so typical of textbooks in philosophy of science in the early 1980s, was uppermost in his mind. This is not so. All he wanted to impress upon his readers was that there is no direct, immediate, empiricist access to nature. As he magnificently put it in a strongly Baconian phrase: "There is a mask of theory over the whole face of nature." This leaves us with the need to demarcate, without the ability to do so. As a historian Whewell decided to live with that dilemma:

> And thus we still have an intelligible distinction of Fact and Theory, if we consider Theory as a conscious, and Fact as an unconscious influence, from the phenomena which are presented to our senses. But still, Theory and Fact, Inference and Perception, Reasoning and Observation, are antitheses in none of which can we separate the two members of a fixed and definite line.

In the history of science, the mode of progress is neither simply cumulative, nor does it proceed by revolutions. If read carefully, the following passage casts a nice balance:

> The principle which constituted the triumph of the preceding stages of the Science, may appear to be subverted and ejected by the

later discoveries, but in fact they are (so far as they were true) taken up into the subsequent doctrines and included in them. They continue to be an essential part of the science. The earlier truths are not expelled but absorbed, not contradicted but extended; and the history of each science which may thus appear like a succession of revolutions, is, in reality, a series of developments."

All through the thousands of pages of the *History* and the *Philosophy*, Whewell explains how concepts change their meaning as they are integrated into new theories. Gravitation, force, attraction, affinity in chemistry, vital forces in physiology; again and again, these change their meaning without forcing the scientist to fall on his logical nose, by telling him that the same concept cannot refer to different entities in different theories in a coherent way. This predicament is circumvented by Whewell's conscious and affirmed acceptance of vague concepts: scientific concepts undergoing change are, perforce, *"vague, but not therefore unmeaning"* (Editor's italics).

What sort of history of science emerges from this view? A unique combination of the old-fashioned and the modern, of the farsighted and the shortsighted; of antipositivist interest in failures and emphasis on the historical role of vague concepts; an evolutionary view of successive developments; a dialectic interaction between facts and ideas. Nevertheless, Whewell unhesitatingly affirms that no real progress had taken place between the Greeks and the age of Copernicus and Galileo. The Arabs, for instance, did not contribute any new knowledge to astronomy. For Whewell, the Middle Ages were indeed the Dark Ages. They are characterized by "Indistinctiveness of Ideas." This is illustrated for several mechanicians, astronomers, and even architects between the time of Archimedes and that of Stevinus or Galileo. On the "positive" side, typical of the Middle Ages was the "commentatorial spirit" and its "natural bias to authority," making authority the dominant source of knowledge of the age.

Such a view would not suggest a detailed interest in medieval mysticism; yet that is just what Whewell has. He describes medieval thinkers in a critical but far from hostile spirit.

> The character of Mysticism is, that it refers to particulars, not to generalisations homogeneous and immediate, but to such as are heterogeneous and remote; to which we must add that the process

of this reference is not a calm act of the intellect, but is accompanied with a glow of enthusiastic feeling.

As a result, Whewell concludes,

> their physical science became Magic, their Astronomy became Astrology, the study of Composition of bodies became Alchemy. Mathematics became the contemplation of the Spiritual Relations of number and figure, and Philosophy became Theosophy.

The Middle Ages were followed by "the dogmatism of the stationary period," typical of scholasticism. In this age, "a universal science was established, with the authority of a religious creed." It is noteworthy that although Whewell saw the period as stationary, and thus incapable of progress, what disturbs him even more is that "error became wicked, dissent became heresy." Yet even this age was not barren of all development:

> How do we recognize, it might be asked, in a picture of mere confusion and mysticism of thought, of senility and dogmatism of character, the powers and requirements to which we owe so many of the most important inventions which we now enjoy?

Whewell deals with this question by distinguishing between Art and Science. Art is practical, Science is speculative. Art is generally prior in time to the related Science. Thus although the inventions made in the Middle Ages became integrated into the sciences, they were originally art, not science; and, indeed, art did progress in the Middle Ages. That such inventions involved an intuitive knowledge of scientific principles does not mean that these were part of science:

> The actions of every man who raises and balances weights or walks along a pole, take for granted the laws of equilibrium; and even animals constantly avail themselves of such principles. Are these, then, acquainted with mechanics as a science?

Science could emerge only when clear ideas again colligated the facts. Therefore, "revived clearness of Ideas" is a prerequisite for the re-emergence of Science, and that occurred only with the revival of letters. This, in Whewell's assessment, is the role of the Renaissance, which also constitutes the Prelude to the Inductive Epoch of Copernicus—the first Inductive Epoch since Hippar-

chus. There is no need to multiply examples. William Whewell—
educator, philosopher, and scientist—was first and foremost a historian; and one very much worth reading and pondering today.

A Note on the Selections

The rationale behind the selections presented here is the wish to
introduce William Whewell, historian, in all his philosophical
complexity.

We start with selections from the two monumental master-
pieces, the basis of Whewell's fame: *History of the Inductive Sciences*
and *The Philosophy of the Inductive Sciences*. The chapters from the
History illustrate Whewell's theory that the growth of science goes
through stages: a Prelude to an Inductive Epoch, the Inductive
Epoch, and the Sequel to it. Thus the Prelude to "the Greek School
of Philosophy" tells us about classical Greece before the Inductive
Epoch of Hipparchus. Here Whewell bravely faces up to the great
historical riddle "Why did the Greeks fail?"

The inductive epoch of Copernicus is preceded by the "mystical" Middle Ages. Whewell the scientist *knows* that there was little
positive addition to scientific knowledge in the Middle Ages, yet
Whewell the historian *knows better* that "positive" knowledge often
grows out of the fertile vagueness of mystical soil. The inductive
epoch of Newton is the pinnacle of scientific achievement. Yet
science in the nineteenth century, though less heroic than in the
age of Newton, made rich contributions to the Newtonian edifice:
thus the chapters on mineralogy and geology.

The chapters from the *Philosophy* illustrate the theme of the
book: Whewell as historian. We can follow his historical treatment
of recurring philosophical issues in the history of scientific thought:
the Fundamental Antithesis of philosophy; the division of the sciences into pure and applied; the historical (Kantian) theory of concepts like Time and Space; and analogy between biology and other
sciences.

The chapters from volume 2 of the *Philosophy* deal with the
historical-cultural determination of knowledge, including inductive knowledge. Whewell's treatment of one of his great heroes,
Francis Bacon, is included here as an important example. In these
chapters Whewell shows how the philosophical problem of the

source of knowledge, empiricism versus innate ideas, is resolved in the historical dialectical process of interaction between the two.

Whewell recast his opinions in the form of aphorisms, following Bacon's view that "this method of presentation was less misleading. It gave a bare outline of . . . discoveries and left obvious blanks when no discoveries had been made. It was a stimulating method which made . . . readers think and judge for themselves" ("Thoughts and Conclusions," in Benjamin Farrington, ed., *The Philosophy of Francis Bacon* [Chicago: University of Chicago Press, 1964], p. 75).

Whewell was much occupied with religious issues and was motivated by his fervent hope that science and religion reinforce each other. This is clearly shown in the chapter from his Bridgewater Treatise, just as the sections from "The Plurality of Worlds" show the direct translation from science to religion and vice versa. These sections, as well as the sections from his essay on induction, show the style of the Whewellian controversies. In his debates with Mill and Brewster, Whewell exhibited clarity of expression, mastery of the facts, and a healthy dose of cunning rhetoric, generally relying on historical rather than logical arguments.

The last essay, "On the Transformations of Hypotheses in the History of Science," is perhaps Whewell's best historical-philosophical analysis. It is a literary gem.

The only change made in these selections has been to provide continuous pagination for this edition. The page numbers in footnotes that refer both to passages in this edition and to others not selected are to the pagination of the original edition.

YEHUDA ELKANA

Bibliographical Note

WORKS BY WHEWELL

1834. *Astronomy and General Physics Considered with Reference to Natural Theology*. Bridgewater Treatises, No. 3, 4th ed. London: William Pickering.

1844. *On the Fundamental Antithesis of Philosophy*, Cambridge Philosophical Society, VII–VIII (1844), 170–181.

[1847] 1967. *The Philosophy of the Inductive Sciences*. In 2 volumes. Facsimilie of 2d ed. New York: Johnson Reprint Corp.

1849. *Of Induction, With Especial Reference to Mr. J. Stuart Mill's System of Logic*. London: John W. Parker.

1854. *Of the Plurality of Worlds: An Essay; also a Dialogue on the Same Subject*. 3d ed. London: John W. Parker.

[1857] 1967. *History of the Inductive Sciences*. In 3 volumes. Reprint of 3d ed. London: Frank Cass & Co.

1866. "Comte and Positivism." *Macmillan's Magazine* 13: 353–62.

1876. *William Whewell: An Account of his Writings, with Selections from his Literary and Scientific Correspondence*, ed. I. Todhunter. In 2 volumes. London: Macmillan and Co.

SECONDARY LITERATURE

Books

Butts, R. ed., *William Whewell's Theory of Scientific Method*. Pittsburgh: University of Pittsburgh Press.

Cannon, Susan Faye. 1978. *Science in Culture: The Early Victorian Period*. New York: Dawson and Science History Publications.

Turner, Frank Miller. 1974. *Between Science and Religion—The Reaction to Scientific Naturalism in Late Victorian England*. New Haven: Yale University Press.

Thackray, Arnold, and Jack Morell. 1981. *Gentlemen of Science: A History of the British Association for the Advancement of Science*. Oxford: Oxford University Press.

Articles

Cannon, Walter F. 1964. "Contribution to Science and Learning." Part 2 of "William Whewell, F.R.S., 1794–1866." *Notes and Records of the Royal Society* 19:168–91.

Brooke, John Hedley. 1977. "Natural Theology and the Plurality of Worlds: Observations on the Brewster-Whewell Debate." *Annals of Science* 34:221–86.

Laudan, L. 1971. "William Whewell on the Consilience of Inductions." *The Monist*. 55:368–91.

Robson, R. 1964. "Academic Life." Part 1 of "William Whewell, F.R.S., 1794–1866." *Notes and Records of the Royal Society* 19.

Ruse, Michael. 1975. "The Scientific Methodology of William Whewell." *Centaurus* 20:227–57.

Strong, E. W. 1956. "William Whewell and John Stuart Mill: Their Controversy about Scientific Knowledge." *Journal of the History of Ideas* 16:209–231.

SELECTED WRITINGS ON THE

HISTORY OF SCIENCE

HISTORY

OF THE

INDUCTIVE SCIENCES,

FROM THE EARLIEST TO THE PRESENT TIME.

By WILLIAM WHEWELL, D.D.,
MASTER OF TRINITY COLLEGE, CAMBRIDGE.

THE THIRD EDITION, WITH ADDITIONS.
IN THREE VOLUMES.

ΛΑΜΠΑΔΙΑ ΕΧΟΝΤΕΣ ΔΙΑΔΩΣΟΥΣΙΝ ΑΛΛΗΛΟΙΣ.

LONDON:
JOHN W. PARKER AND SON, WEST STRAND.
1857.

INTRODUCTION.

IT is my purpose to write the History of some of the most important of the Physical Sciences, from the earliest to the most recent periods. I shall thus have to trace some of the most remarkable branches of human knowledge, from their first germ to their growth into a vast and varied assemblage of undisputed truths; from the acute, but fruitless, essays of the early Greek Philosophy, to the comprehensive systems, and demonstrated generalizations, which compose such sciences as the Mechanics, Astronomy, and Chemistry, of modern times.

The completeness of historical view which belongs to such a design, consists, not in accumulating all the details of the cultivation of each science, but in marking the larger features of its formation. The historian must endeavour to point out how each of the important advances was made, by which the sciences have reached their present position; and when and by whom each of the valuable truths was obtained, of which the aggregate now constitutes a costly treasure.

Such a task, if fitly executed, must have a well-founded interest for all those who look at the existing condition of human knowledge with complacency and admiration. The present generation finds itself the heir of a vast patrimony of science; and it must needs concern us to know the steps by which these possessions were acquired, and the documents by which they are secured to us and our heirs for ever. Our species, from the time of its creation, has been travelling onwards in pursuit of truth; and now that we have reached a lofty and commanding position, with the broad light of day around us, it must be grateful to look back on the line of our past progress;—to review the journey, begun in early twilight amid primeval wilds; for a long time continued with slow advance and obscure prospects; and gradually and in later days followed along more open and lightsome paths, in a wide and fertile region. The historian of science, from early periods to the present times, may hope for favour on the score of the mere subject of his narrative, and in virtue of the curiosity which the men of the present day may naturally feel respecting the events and persons of his story.

But such a survey may possess also an interest of another kind; it may be instructive as well as agreeable; it may bring before the reader the present form and extent, the future hopes and prospects of science, as well as its past progress. The eminence on which we stand may enable us to see the land of promise, as well as the wilderness through which we have passed. The examination of the steps by which our ancestors acquired our intellectual estate, may make us acquainted with our expectations as well as our possessions;—may not only remind us of what we have, but may teach us how to improve and increase our store. It will be universally expected that a History of Inductive Science should point out to us a philosophical distribution of the existing body of knowledge, and afford us some indication of the most promising mode of directing our future efforts to add to its extent and completeness.

To deduce such lessons from the past history of human knowledge, was the intention which originally gave rise to the present work. Nor is this portion of the design in any measure abandoned; but its execution, if it take place, must be attempted in a separate and future treatise, *On the Philosophy of the Inductive Sciences.* An essay of this kind may, I trust, from the progress already made in it, be laid before the public at no long interval after the present history.[1]

Though, therefore, many of the principles and maxims

---

[1] The *Philosophy of the Inductive Sciences* was published shortly after the present work.

of such a work will disclose themselves with more or less of distinctness in the course of the history on which we are about to enter, the systematic and complete exposition of such principles must be reserved for this other treatise. My attempts and reflections have led me to the opinion that justice cannot be done to the subject without such a division of it.

To this future work, then, I must refer the reader who is disposed to require, at the outset, a precise explanation of the terms which occur in my title. It is not possible, without entering into this philosophy, to explain adequately how science which is INDUCTIVE differs from that which is not so; or why some portions of *knowledge* may properly be selected from the general mass and termed SCIENCE. It will be sufficient at present to say, that the sciences of which we have here to treat, are those which are commonly known as the *Physical Sciences;* and that by *Induction* is to be understood that process of collecting general truths from the examination of particular facts, by which such sciences have been formed.

There are, however, two or three remarks, of which the application will occur so frequently, and will tend so much to give us a clearer view of some of the subjects which occur in our history, that I will state them now in a brief and general manner.

Facts and Ideas.[2]—In the first place, then, I remark, that, to the formation of science, two things are requisite; —Facts and Ideas; observation of Things without, and an inward effort of Thought; or, in other words, Sense and Reason. Neither of these elements, by itself, can constitute substantial general knowledge. The impressions of sense, unconnected by some rational and speculative principle, can only end in a practical acquaintance with individual objects; the operations of the rational faculties, on the other hand, if allowed to go on without a constant reference to external things, can lead only to empty abstraction and barren ingenuity.

---

[2] For the *Antithesis of Facts and Ideas*, see the *Philosophy*, book i. ch. 1, 2, 4, 5.

Real speculative knowledge demands the combination of the two ingredients;—right reason, and facts to reason upon. It has been well said, that true knowledge is the interpretation of nature; and therefore it requires both the interpreting mind, and nature for its subject; both the document, and the ingenuity to read it aright. Thus invention, acuteness, and connexion of thought, are necessary on the one hand, for the progress of philosophical knowledge; and on the other hand, the precise and steady application of these faculties to facts well known and clearly conceived. It is easy to point out instances in which science has failed to advance, in consequence of the absence of one or other of these requisites; indeed, by far the greater part of the course of the world, the history of most times and most countries, exhibits a condition thus stationary with respect to knowledge. The facts, the impressions on the senses, on which the first successful attempts at physical knowledge proceeded, were as well known long before the time when they were thus turned to account, as at that period. The motions of the stars, and the effects of weight, were familiar to man before the rise of the Greek Astronomy and Mechanics: but the " diviner mind" was still absent; the act of thought had not been exerted, by which these facts were bound together under the form of laws and principles. And even at this day, the tribes of uncivilized and half-civilized man over the whole face of the earth, have before their eyes a vast body of facts, of exactly the same nature as those with which Europe has built the stately fabric of her physical philosophy; but, in almost every other part of the earth, the process of the intellect by which these facts become science, is unknown. The scientific faculty does not work. The scattered stones are there, but the builder's hand is wanting. And again, we have no lack of proof that mere activity of thought is equally inefficient in producing real knowledge. Almost the whole of the career of the Greek schools of philosophy; of the schoolmen of Europe in the middle ages; of the Arabian and Indian philosophers; shows us that we may have extreme ingenuity

INTRODUCTION.

and subtlety, invention and connexion, demonstration and method; and yet that out of these germs, no physical science may be developed. We may obtain, by such means, Logic and Metaphysics, and even Geometry and Algebra; but out of such materials we shall never form Mechanics and Optics, Chemistry and Physiology. How impossible the formation of these sciences is without a constant and careful reference to observation and experiment;—how rapid and prosperous their progress may be when they draw from such sources the materials on which the mind of the philosopher employs itself;—the history of those branches of knowledge for the last three hundred years abundantly teaches us.

Accordingly, the existence of clear Ideas applied to distinct Facts will be discernible in the History of Science, whenever any marked advance takes place. And, in tracing the progress of the various provinces of knowledge which come under our survey, it will be important for us to see that, at all such epochs, such a combination has occurred; that whenever any material step in general knowledge has been made,—whenever any philosophical discovery arrests our attention,—some man or men come before us, who have possessed, in an eminent degree, a clearness of the ideas which belong to the subject in question, and who have applied such ideas in a vigorous and distinct manner to ascertained facts and exact observations. We shall never proceed through any considerable range of our narrative, without having occasion to remind the reader of this reflection.

Successive Steps in Science.[3]—But there is another remark which we must also make. Such sciences as we have here to do with are, commonly, not formed by a single act;—they are not completed by the discovery of one great principle. On the contrary, they consist in a long-continued advance; a series of changes; a repeated progress from one principle to another, diffe-

---

[3] Concerning *Successive Generalizations in Science,* see the *Philosophy,* book i. ch. 2, sect. 11.

7

rent and often apparently contradictory. Now, it is important to remember that this contradiction is apparent only. The principles which constituted the triumph of the preceding stages of the science, may appear to be subverted and ejected by the later discoveries, but in fact they are, (so far as they were true,) taken up into the subsequent doctrines and included in them. They continue to be an essential part of the science. The earlier truths are not expelled but absorbed, not contradicted but extended; and the history of each science, which may thus appear like a succession of revolutions, is, in reality, a series of developments. In the intellectual, as in the material world,—

> Omnia mutantur nil interit . . . . .
> Nec manet ut fuerat nec formas servat easdem,
> Sed tamen ipsa eadem est.
>
> All changes, nought is lost; the forms are changed,
> And that which has been is not what it was,
> Yet that which has been is.

Nothing which was done was useless or unessential, though it ceases to be conspicuous and primary.

Thus the final form of each science contains the substance of each of its preceding modifications; and all that was at any antecedent period discovered and established, ministers to the ultimate development of its proper branch of knowledge. Such previous doctrines may require to be made precise and definite, to have their superfluous and arbitrary portions expunged, to be expressed in new language, to be taken up into the body of science by various processes;—but they do not on such accounts cease to be true doctrines, or to form a portion of the essential constituents of our knowledge.

Terms record Discoveries.[4]—The modes in which the earlier truths of science are preserved in its later forms, are indeed various. From being asserted at first as strange discoveries, such truths come at last to be implied as almost self-evident axioms. They are recorded

---

[4] Concerning *Technical Terms*, see *Philosophy*, book i. ch. 3.

8

by some familiar maxim, or perhaps by some new word or phrase, which becomes part of the current language of the philosophical world; and thus asserts a principle, while it appears merely to indicate a transient notion;— preserves as well as expresses a truth;—and, like a medal of gold, is a treasure as well as a token. We shall frequently have to notice the manner in which great discoveries thus stamp their impress upon the terms of a science; and, like great political revolutions, are recorded by the change of the current coin which has accompanied them.

Generalization.—The great changes which thus take place in the history of science, the revolutions of the intellectual world, have, as a usual and leading character, this, that they are steps of *generalization;*—transitions from particular truths to others of a wider extent, in which the former are included. This progress of knowledge, from individual facts to universal laws,—from particular propositions to general ones,—and from these to others still more general, with reference to which the former generalizations are particular,—is so far familiar to men's minds, that, without here entering into further explanation, its nature will be understood sufficiently to prepare the reader to recognise the exemplifications of such a process, which he will find at every step of our advance.

Inductive Epochs; Preludes; Sequels.—In our history, it is the *progress* of knowledge only which we have to attend to. This is the main action of our drama; and all the events which do not bear upon this, though they may relate to the cultivation and the cultivators of philosophy, are not a necessary part of our theme. Our narrative will therefore consist mainly of successive steps of generalization, such as have just been mentioned. But among these, we shall find some of eminent and decisive importance, which have more peculiarly influenced the fortunes of physical philosophy, and to which we may consider the rest as subordinate and auxiliary. These primary movements, when the Inductive process, by which science is formed, has been exercised in a more energetic and powerful manner, may be distinguished as the *Inductive Epochs* of scientific history; and they deserve our more express and pointed notice. They are, for the most part, marked by the great discoveries and the great philosophical names which all civilized nations have agreed in admiring. But, when we examine more clearly the history of such discoveries, we find that these epochs have not occurred suddenly and without preparation. They have been preceded by a period, which we may call their *Prelude*, during which the ideas and facts on which they turned were called into action;—were gradually evolved into clearness and connexion, permanency and certainty; till at last the discovery which marks the Epoch, seized and fixed for ever the truth which had till then been obscurely and doubtfully discerned. And again, when this step has been made by the principal discoverers, there may generally be observed another period, which we may call the *Sequel* of the Epoch, during which the discovery has acquired a more perfect certainty and a more complete development among the leaders of the advance; has been diffused to the wider throng of the secondary cultivators of such knowledge, and traced into its distant consequences. This is a work, always of time and labour, often of difficulty and conflict. To distribute the History of science into such Epochs, with their Preludes and Sequels, if successfully attempted, must needs make the series and connexions of its occurrences more distinct and intelligible. Such periods form resting-places, where we pause till the dust of the confused march is laid, and the prospect of the path is clear.

Inductive Charts.[5]—Since the advance of science consists in collecting by induction true general laws from particular facts, and in combining several such laws into one higher generalization, in which they still retain their truth; we might form a Chart, or Table, of the progress of each science, by setting down the particular facts which have thus been combined, so as

---

[5] Inductive charts of the History of Astronomy and of Optics, such as are here referred to, are given in the *Philosophy*, book xi. ch. 6.

to form general truths, and by marking the further union of these general truths into others more comprehensive. The Table of the progress of any science would thus resemble the Map of a River, in which the waters from separate sources unite and make rivulets, which again meet with rivulets from other fountains, and thus go on forming by their junction trunks of a higher and higher order. The representation of the state of a science in this form, would necessarily exhibit all the principal doctrines of the science; for each general truth contains the particular truths from which it was derived, and may be followed backwards till we have these before us in their separate state. And the last and most advanced generalization would have, in such a scheme, its proper place and the evidence of its validity. Hence such an *Inductive Table* of each science would afford a criterion of the correctness of our distribution of the inductive Epochs, by its coincidence with the views of the best judges, as to the substantial contents of the science in question. By forming, therefore, such Inductive Tables of the principal sciences of which I have here to speak, and by regulating by these tables, my views of the history of the sciences, I conceive that I have secured the distribution of my history from material error; for no merely arbitrary division of the events could satisfy such conditions. But though I have constructed such charts to direct the course of the present history, I shall not insert them in the work, reserving them for the illustration of the philosophy of the subject; for to this they more properly belong, being a part of the *Logic of Induction.*

Stationary Periods.—By the lines of such maps the real advance of science is depicted, and nothing else. But there are several occurrences of other kinds, too interesting and too instructive to be altogether omitted. In order to understand the conditions of the progress of knowledge, we must attend, in some measure, to the failures as well as the successes by which such attempts have been attended. When we reflect during how small a portion of the whole history of human speculations, science has really been, in any marked degree, progressive, we must needs feel some curiosity to know what was doing in these *stationary* periods; what field could be found which admitted of so wide a deviation, or at least so protracted a wandering. It is highly necessary to our purpose, to describe the baffled enterprises as well as the achievements of human speculation.

Deduction.—During a great part of such stationary periods, we shall find that the process which we have spoken of as essential to the formation of real science, the conjunction of clear Ideas with distinct Facts, was interrupted; and, in such cases, men dealt with ideas alone. They employed themselves in reasoning from principles, and they arranged, and classified, and analyzed their ideas, so as to make their reasonings satisfy the requisitions of our rational faculties. This process of drawing conclusions from our principles, by rigorous and unimpeachable trains of demonstration, is termed *Deduction.* In its due place, it is a highly important part of every science; but it has no value when the fundamental principles, on which the whole of the demonstration rests, have not first been obtained by the induction of facts, so as to supply the materials of substantial truth. Without such materials, a series of demonstrations resembles physical science only as a shadow resembles a real object. To give a real significance to our propositions, Induction must provide what Deduction cannot supply. From a pictured hook we can hang only a pictured chain.

Distinction of common Notions and Scientific Ideas.[6] —When the notions with which men are conversant in the common course of practical life, which give meaning to their familiar language, and employment to their hourly thoughts, are compared with the Ideas on which exact science is founded, we find that the two classes of intellectual operations have much that is common and much that is different. Without here attempting fully to explain this relation, (which, indeed, is one

---

[6] Scientific Ideas depend upon certain *Fundamental Ideas*, which are enumerated in the *Philosophy*, book i. ch. 8.

of the hardest problems of our philosophy,) we may observe that they have this in common, that both are acquired by acts of the mind exercised in connecting external impressions, and may be employed in conducting a train of reasoning; or, speaking loosely, (for we cannot here pursue the subject so as to arrive at philosophical exactness,) we may say, that all notions and ideas are obtained by an *inductive,* and may be used in a *deductive* process. But scientific Ideas and common Notions differ in this, that the former are precise and stable, the latter vague and variable; the former are possessed with clear insight, and employed in a sense rigorously limited, and always identically the same; the latter have grown up in the mind from a thousand dim and diverse suggestions, and the obscurity and incongruity which belong to their origin hang about all their applications. Scientific Ideas can often be adequately exhibited for all the purposes of reasoning, by means of Definitions and Axioms; all attempts to reason by means of Definitions from common Notions, lead to empty forms or entire confusion.

Such common Notions are sufficient for the common practical conduct of human life; but man is not a practical creature merely; he has within him a *speculative* tendency, a pleasure in the contemplation of ideal relations, a love of knowledge *as* knowledge. It is this speculative tendency which brings to light the difference of common Notions and scientific Ideas, of which we have spoken. The mind analyzes such Notions, reasons upon them, combines and connects them; for it feels assured that intellectual things ought to be able to bear such handling. Even practical knowledge, we see clearly, is not possible without the use of the reason; and the speculative reason is only the reason satisfying itself of its own consistency. This speculative faculty cannot be controlled from acting. The mind cannot but claim a right to speculate concerning all its own acts and creations; yet, when it exercises this right upon its common practical notions, we find that it runs into barren abstractions and ever-recurring cycles of subtlety. Such Notions are like waters naturally stagnant; however much we urge and agitate them, they only revolve in stationary whirlpools. But the mind is capable of acquiring scientific Ideas, which are better fitted to undergo discussion and impulsion. When our speculations are duly fed from the spring-heads of Observation, and frequently drawn off into the region of Applied Science, we may have a living stream of consistent and progressive knowledge. That science may be both real as to its import, and logical as to its form, the examples of many existing sciences sufficiently prove.

School Philosophy.—So long, however, as attempts are made to form sciences, without such a verification and realization of their fundamental ideas, there is, in the natural series of speculation, no self-correcting principle. A philosophy constructed on notions obscure, vague, and unsubstantial, and held in spite of the want of correspondence between its doctrines and the actual train of physical events, may long subsist, and occupy men's minds. Such a philosophy must depend for its permanence upon the pleasure which men feel in tracing the operations of their own and other men's minds, and in reducing them to logical consistency and systematical arrangement.

In these cases the main subjects of attention are not external objects, but speculations previously delivered; the object is not to interpret nature, but man's mind. The opinions of the Masters are the facts which the Disciples endeavour to reduce to unity, or to follow into consequences. A series of speculators who pursue such a course, may properly be termed a *School*, and their philosophy a *School Philosophy;* whether their agreement in such a mode of seeking knowledge arise from personal communication and tradition, or be merely the result of a community of intellectual character and propensity. The two great periods of School Philosophy (it will be recollected that we are here directing our attention mainly to physical science), were that of the Greeks and that of the Middle Ages;—the period of the first waking of science, and that of its mid-day slumber.

What has been said thus briefly and imperfectly, would require great detail and much explanation, to give it its full significance and authority. But it seemed proper to state so much in this place, in order to render more intelligible and more instructive at the first aspect, the view of the attempted or effected progress of science.

It is, perhaps, a disadvantage inevitably attending an undertaking like the present, that it must set out with statements so abstract; and must present them without their adequate development and proof. Such an Introduction, both in its character and its scale of execution, may be compared to the geographical sketch of a country, with which the historian of its fortunes often begins his narration. So much of Metaphysics is as necessary to us as such a portion of Geography is to the Historian of an Empire; and what has hitherto been said, is intended as a slight outline of the Geography of that Intellectual World, of which we have here to study the History.

The name which we have given to this History— A HISTORY OF THE INDUCTIVE SCIENCES—has the fault of seeming to exclude from the rank of Inductive Sciences those which are not included in the History; as Ethnology and Glossology, Political Economy, Psychology. This exclusion I by no means wish to imply; but I could find no other way of compendiously describing my subject, which was intended to comprehend those Sciences in which, by the observation of facts and the use of reason, systems of doctrine have been established which are universally received as truths among thoughtful men; and which may therefore be studied as examples of the manner in which truth is to be discovered. Perhaps a more exact description of the work would have been, *A History of the principal Sciences hitherto established by Induction.* I may add that I do not include in the phrase 'Inductive Sciences,' the branches of Pure Mathematics, (Geometry, Arithmetic, Algebra, and the like,) because, as I have elsewhere stated (*Phil. Ind. Sc.*, book ii. c. 1), these are not *Inductive* but *Deductive* Sciences. They do not infer true theories

15

from observed facts, and more general from more limited laws: but they trace the conditions of all theory, the properties of space and number; and deduce results from ideas without the aid of experience. The History of these Sciences is briefly given in Chapters 13 and 14 of the Second Book of the *Philosophy* just referred to.

I may further add that the other work to which I refer, *the Philosophy of the Inductive Sciences*, is in a great measure historical, no less than the present *History*. That work contains the history of the Sciences so far as it depends on *Ideas;* the present work contains the history so far as it depends upon *Observation*. The two works resulted simultaneously from the same examination of the principal writers on science in all ages, and may serve to supplement each other.

BOOK I.

HISTORY OF THE GREEK SCHOOL PHILOSOPHY,
WITH REFERENCE TO PHYSICAL SCIENCE.

CHAPTER I.

PRELUDE TO THE GREEK SCHOOL PHILOSOPHY.

Sect. 1.—*First Attempts of the Speculative Faculty in
Physical Inquiries.*

AT an early period of history there appeared in men
a propensity to pursue speculative inquiries concerning the various parts and properties of the material
world. What they saw excited them to meditate, to
conjecture, and to reason: they endeavoured to account
for natural events, to trace their causes, to reduce
them to their principles. This habit of mind, or, at
least that modification of it which we have here to
consider, seems to have been first unfolded among the
Greeks. And during that obscure introductory interval
which elapsed while the speculative tendencies of men
were as yet hardly disentangled from the practical,
those who were most eminent in such inquiries were
distinguished by the same term of praise which is applied to sagacity in matters of action, and were called
wise men—σοφοί. But when it came to be clearly felt
by such persons that their endeavours were suggested
by the love of knowledge, a motive different from the
motives which lead to the wisdom of active life, a name
was adopted of a more appropriate, as well as of a more
modest signification, and they were termed *philosophers,*
or lovers of wisdom. This appellation is said[1] to have
been first assumed by Pythagoras. Yet he, in Hero-

---

[1] Cic. Tusc. v. 3.

dotus, instead of having this title, is called a powerful
sophist—Ἑλλήνων οὐ τῷ ἀσθενεστάτῳ σοφιστῇ Πυθαγόρῃ;[2]
the historian using this word, as it would seem, without
intending to imply that misuse of reason which the
term afterwards came to denote. The historians of
literature placed Pythagoras at the origin of the Italic
School, one of the two main lines of succession of the
early Greek philosophers: but the other, the Ionic
School, which more peculiarly demands our attention,
in consequence of its character and subsequent progress,
is deduced from Thales, who preceded the age of *Philosophy*, and was one of the *sophi*, or 'wise men of
Greece.'

The Ionic School was succeeded in Greece by several
others; and the subjects which occupied the attention
of these schools became very extensive. In fact, the
first attempts were, to form systems which should explain the laws and causes of the material universe; and
to these were soon added all the great questions which
our moral condition and faculties suggest. The physical
philosophy of these schools is especially deserving of our
study, as exhibiting the character and fortunes of the
most memorable attempt at universal knowledge which
has ever been made. It is highly instructive to trace
the principles of this undertaking; for the course
pursued was certainly one of the most natural and
tempting which can be imagined; the essay was made
by a nation unequalled in fine mental endowments, at
the period of its greatest activity and vigour; and yet
it must be allowed, (for, at least so far as physical
science is concerned, none will contest this,) to have
been entirely unsuccessful. We cannot consider otherwise than as an utter failure, an endeavour to discover
the causes of things, of which the most complete results
are the Aristotelian physical treatises; and which, after
reaching the point which these treatises mark, left the
human mind to remain stationary, at any rate on all
such subjects, for nearly two thousand years.

The early philosophers of Greece entered upon the

---

2 Herod. iv. 95.

work of physical speculation in a manner which showed the vigour and confidence of the questioning spirit, as yet untamed by labours and reverses. It was for later ages to learn that man must acquire, slowly and patiently, letter by letter, the alphabet in which nature writes her answers to such inquiries. The first students wished to divine, at a single glance, the whole import of her book. They endeavoured to discover the origin and principle of the universe; according to Thales, *water* was the origin of all things, according to Anaximenes, *air;* and Heraclitus considered *fire* as the essential principle of the universe. It has been conjectured, with great plausibility, that this tendency to give to their Philosophy the form of a Cosmogony, was owing to the influence of the poetical Cosmogonies and Theogonies which had been produced and admired at a still earlier age. Indeed, such wide and ambitious doctrines as those which have been mentioned, were better suited to the dim magnificence of poetry, than to the purpose of a philosophy which was to bear the sharp scrutiny of reason. When we speak of the *principles* of things, the term, even now, is very ambiguous and indefinite in its import, but how much more was that the case in the first attempts to use such abstractions ! The term which is commonly used in this sense (ἀρχή), signified at first *the beginning;* and in its early philosophical applications implied some obscure mixed reference to the mechanical, chemical, organic, and historical causes of the visible state of things, besides the theological views which at this period were only just beginning to be separated from the physical. Hence we are not to be surprised if the sources from which the opinions of this period appear to be derived are rather vague suggestions and casual analogies, than any reasons which will bear examination. Aristotle conjectures, with considerable probability, that the doctrine of Thales, according to which water was the universal element, resulted from the manifest importance of moisture in the support of animal and vegetable life.[3] But such

---

[3] Metaph. i. 3.

precarious analyses of these obscure and loose dogmas of early antiquity are of small consequence to our object.

In more limited and more definite examples of inquiry concerning the causes of natural appearances, and in the attempts made to satisfy men's curiosity in such cases, we appear to discern a more genuine prelude to the true spirit of physical inquiry. One of the most remarkable instances of this kind is to be found in the speculations which Herodotus records, relative to the cause of the floods of the Nile. 'Concerning the nature of this river,' says the father of history,[4] 'I was not able to learn anything, either from the priests or from any one besides, though I questioned them very pressingly. For the Nile is flooded for a hundred days, beginning with the summer solstice; and after this time it diminishes, and is, during the whole winter, very small. And on this head I was not able to obtain anything satisfactory from any one of the Egyptians, when I asked what is the power by which the Nile is in its nature the reverse of other rivers.'

We may see, I think, in the historian's account, that the Grecian mind felt a craving to discover the reasons of things which other nations did not feel. The Egyptians, it appears, had no theory, and felt no want of a theory. Not so the Greeks; they had their reasons to render, though they were not such as satisfied Herodotus. 'Some of the Greeks,' he says, 'who wish to be considered great philosophers, (Ἑλλήνων τινες επισήμοι βουλόμενοι γενέσθαι σοφίην) have propounded three ways of accounting for these floods. Two of them,' he adds, 'I do not think worthy of record, except just so far as to mention them.' But as these are some of the earliest Greek essays in physical philosophy, it will be worth while, even at this day, to preserve the brief notice he has given of them, and his own reasonings upon the same subject.

'One of these opinions holds that the Etesian winds [which blew from the north] are the cause of these

---

[4] Herod. ii. 19.

PRELUDE.

floods, by preventing the Nile from flowing into the
sea.' Against this the historian reasons very simply
and sensibly. 'Very often when the Etesian winds do
not blow, the Nile is flooded nevertheless. And moreover, if the Etesian winds were the cause, all other
rivers, which have their course opposite to these winds,
ought to undergo the same changes as the Nile; which
the rivers of Syria and Libya so circumstanced do
not.'

'The next opinion is still more unscientific, (ἀνεπι-
στημονεστέρη) and is, in truth, marvellous for its folly.
This holds that the ocean flows all round the earth, and
that the Nile comes out of the ocean, and by that means
produces its effects.' 'Now,' says the historian, 'the
man who talks about this ocean-river, goes into the
region of fable, where it is not easy to demonstrate that
he is wrong. I know of no such river. But I suppose
that Homer or some of the earlier poets invented this
fiction and introduced it into their poetry.'

He then proceeds to a third account, which to a
modern reasoner would appear not at all unphilosophical in itself, but which he, nevertheless, rejects in
a manner no less decided than the others. 'The third
opinion, though much the most plausible, is still more
wrong than the others; for it asserts an impossibility,
namely, that the Nile proceeds from the melting of the
snow. Now the Nile flows out of Libya, and through
Ethiopia, which are very hot countries, and thus comes
into Egypt, which is a colder region. How then can
it proceed from snow?' He then offers several other
reasons 'to show,' as he says, 'to any one capable of
reasoning on such subjects' (ἀνδρί γε λογίζεσθαι τοιούτων
πέρι οἵῳ τε ἐόντι), that the assertion cannot be true.
The winds which blow from the southern regions are
hot; the inhabitants are black; the swallows and kites
(ἰκτῖνοι) stay in the country the whole year; the cranes
fly the colds of Scythia, and seek their warm winter-
quarters there; which would not be if it snowed ever
so little.' He adds another reason, founded apparently
upon some limited empirical maxim of weather-wisdom
taken from the climate of Greece. 'Libya,' he said,

21

'has neither rain nor ice, and therefore no snow; *for*, in five days after a fall of snow there must be a fall of rain; so that if it snowed in those regions it must rain too.' I need not observe that Herodotus was not aware of the difference between the climate of high mountains and plains in a torrid region; but it is impossible not to be struck both with the activity and the coherency of thought displayed by the Greek mind in this primitive physical inquiry.

But I must not omit the hypothesis which Herodotus himself proposes, after rejecting those which have been already given. It does not appear to me easy to catch his exact meaning, but the statement will still be curious. 'If,' he says, 'one who has condemned opinions previously promulgated may put forwards his own opinion concerning so obscure a matter, I will state why it seems to me that the Nile is flooded in summer.' This opinion he propounds at first with an oracular brevity, which it is difficult to suppose that he did not intend to be impressive. 'In winter the sun is carried by the seasons away from his former course, and goes to the upper parts of Libya. And *there, in short, is the whole account;* for that region to which this divinity (the sun) is nearest, must naturally be most scant of water, and the river-sources of that country must be dried up.'

But the lively and garrulous Ionian immediately relaxes from this apparent reserve. 'To explain the matter more at length,' he proceeds, 'it is thus. The sun, when he traverses the upper parts of Libya, does what he commonly does in summer;—he *draws* the water to him (ἕλκει ἐπ᾽ ἑωϋτὸν τὸ ὕδωρ), and having thus drawn it, he pushes it to the upper regions (of the air probably,) and then the winds take it and disperse it till they dissolve in moisture. And thus the winds which blow from those countries, Libs and Notus, are the most moist of all winds. Now when the winter relaxes and the sun returns to the north, he still draws water from all the rivers, but they are increased by showers and rain-torrents, so that they are in flood till the summer comes; and then, the rain failing and the sun still drawing them, they become small. But the Nile, not being fed by rains, yet being drawn by the sun, is, alone of all rivers, much more scanty in the winter than in the summer. For in summer it is drawn like all other rivers, but in winter it alone has its supplies shut up. And in this way, I have been led to think the sun is the cause of the occurrence in question.' We may remark that the historian here appears to ascribe the inequality of the Nile at different seasons to the influence of the sun upon its springs alone, the other cause of change, the rains, being here excluded: and that, on this supposition, the same relative effects would be produced whether the sun increase the sources in winter by melting the snows, or diminish them in summer by what he calls *drawing* them upwards.

This specimen of the early efforts of the Greeks in physical speculations, appears to me to speak strongly for the opinion that their philosophy on such subjects was the native growth of the Greek mind, and owed nothing to the supposed lore of Egypt and the East; an opinion which has been adopted with regard to the Greek philosophy in general by the most competent judges on a full survey of the evidence.[5] Indeed, we have no evidence whatever that, at any period, the African or Asiatic nations, (with the exception perhaps of the Indians,) ever felt this importunate curiosity with regard to the definite application of the idea of cause and effect to visible phenomena; or drew so strong a line between a fabulous legend and a reason rendered; or attempted to ascend to a natural cause by classing together phenomena of the same kind. We may be well excused, therefore, for believing that they could not impart to the Greeks what they themselves did not possess; and so far as our survey goes, physical philosophy has its origin, apparently spontaneous and independent, in the active and acute intellect of Greece.

---

[5] Thirlwall, *Hist. Gr.*, ii. 130; and, as there quoted, Ritter, *Geschichte der Philosophie*, i. 159—173.

Sect. 2.—*Primitive Mistake in Greek Physical Philosophy.*

WE now proceed to examine with what success the Greeks followed the track into which they had thus struck. And here we are obliged to confess that they very soon turned aside from the right road to truth, and deviated into a vast field of error, in which they and their successors have wandered almost to the present time. It is not necessary here to inquire why those faculties which appear to be bestowed upon us for the discovery of truth, were permitted by Providence to fail so signally in answering that purpose; whether, like the powers by which we seek our happiness, they involve a responsibility on our part, and may be defeated by rejecting the guidance of a higher faculty; or whether these endowments, though they did not immediately lead man to profound physical knowledge, answered some nobler and better purpose in his constitution and government. The fact undoubtedly was, that the physical philosophy of the Greeks soon became trifling and worthless; and it is proper to point out, as precisely as we can, in what the fundamental mistake consisted.

To explain this, we may in the first place return for a moment to Herodotus's account of the cause of the floods of the Nile.

The reader will probably have observed a remarkable phrase used by Herodotus, in his own explanation of these inundations. He says that the sun *draws*, or attracts, the water; a metaphorical term, obviously intended to denote some more general and abstract conception than that of the visible operation which the word primarily signifies. This abstract notion of 'drawing' is, in the historian, as we see, very vague and loose; it might, with equal propriety, be explained to mean what we now understand by mechanical or by chemical attraction, or pressure, or evaporation. And in like manner, all the first attempts to comprehend the operations of nature, led to the introduction of abstract conceptions, often vague, indeed, but not, therefore, unmeaning; such as *motion* and *velocity*, *force* and *pressure*, *impetus* and *momentum* (ῥοπή). And the next step in philosophizing, necessarily was to endeavour to make these vague abstractions more clear and fixed, so that the logical faculty should be able to employ them securely and coherently. But there were two ways of making this attempt; the one, by examining the words only, and the thoughts which they call up; the other, by attending to the facts and things which bring these abstract terms into use. The latter, the method of *real* inquiry, was the way to success; but the Greeks followed the former, the *verbal* or *notional* course, and failed.

If Herodotus, when the notion of the sun's attracting the waters of rivers had entered into his mind, had gone on to instruct himself, by attention to facts, in what manner this notion could be made more definite, while it still remained applicable to all the knowledge which could be obtained, he would have made some progress towards a true solution of his problem. If, for instance, he had tried to ascertain whether this Attraction which the sun exerted upon the waters of rivers, depended on his influence at their fountains only, or was exerted over their whole course, and over waters which were not parts of rivers, he would have been led to reject his hypothesis; for he would have found, by observations sufficiently obvious, that the sun's Attraction, as shown in such cases, is a tendency to lessen all expanded and open collections of moisture, whether flowing from a spring or not; and it would then be seen that this influence, operating on the whole surface of the Nile, must diminish it as well as other rivers, in summer, and therefore could not be the cause of its overflow. He would thus have corrected his first loose conjecture by a real study of nature, and might, in the course of his meditations, have been led to available notions of Evaporation, or other natural actions. And, in like manner, in other cases, the rude attempts at explanation, which the first exercise of the speculative faculty produced, might have been gradually concentrated and refined, so as to fall in, both with the requisitions of reason and the testimony of sense.

But this was not the direction which the Greek speculators took. On the contrary; as soon as they had introduced into their philosophy any abstract and general conceptions, they proceeded to scrutinize these by the internal light of the mind alone, without any longer looking abroad into the world of sense. They took for granted that philosophy must result from the relations of those notions which are involved in the common use of language, and they proceeded to seek their philosophical doctrines by studying such notions. They ought to have reformed and fixed their usual conceptions by Observation; they only analysed and expanded them by Reflection: they ought to have sought by trial, among the Notions which passed through their minds, some one which admitted of exact application to Facts; they selected arbitrarily, and, consequently, erroneously, the Notions according to which Facts should be assembled and arranged: they ought to have collected clear Fundamental Ideas from the world of things by *inductive* acts of thought; they only derived results by *Deduction* from one or other of their familiar Conceptions.[6]

When this false direction had been extensively adopted by the Greek philosophers, we may treat of it as the method of their *Schools*. Under that title we must give a further account of it.

---

[6] The course by which the Sciences were formed, and which is here referred to as that which the Greeks did *not* follow, is described in detail in the *Philosophy*, book xi., *Of the Construction of Science.*

CHAPTER III.

OF THE MYSTICISM OF THE MIDDLE AGES.

IT has been already several times hinted, that a new and peculiar element was introduced into the Greek philosophy which occupied the attention of the Alexandrian school; and that this element tinged a large portion of the speculations of succeeding ages. We may speak of this peculiar element as *Mysticism;* for, from the notion usually conveyed by this term, the reader will easily apprehend the general character of the tendency now spoken of; and especially when he sees its effect pointed out in various subjects. Thus, instead of referring the events of the external world to space and time, to sensible connexion and causation, men attempted to reduce such occurrences under spiritual and supersensual relations and dependencies; they referred them to superior intelligences, to theological conditions, to past and future events in the moral world, to states of mind and feelings, to the creatures of an imaginary mythology or demonology. And thus their physical Science became Magic, their Astronomy became Astrology, the study of the Composition of bodies became Alchemy, Mathematics became the contemplation of the Spiritual Relations of number and figure, and Philosophy became Theosophy.

The examination of this feature in the history of the human mind is important for us, in consequence of its influence upon the employments and the thoughts of the times now under our notice. This tendency materially affected both men's speculations and their labours in the pursuit of knowledge. By its direct operation, it gave rise to the newer Platonic philosophy among the Greeks, and to corresponding doctrines among the Arabians; and by calling into a prominent place astrology, alchemy, and magic, it long occupied most of the real observers of the material world. In this manner it delayed and impeded the progress of true science; for we shall see reason to believe that human knowledge lost more by the perversion of men's minds and the misdirection of their efforts, than it gained by any increase of zeal arising from the peculiar hopes and objects of the mystics.

It is not to our purpose to attempt any general view of the progress and fortunes of the various forms of Mystical Philosophy; but only to exhibit some of its characters, in so far as they illustrate those tendencies of thought which accompanied the retrogradation of inductive science. And of these, the leading feature which demands our notice is that already alluded to; namely, the practice of referring things and events, not to clear and distinct relations, obviously applicable to such cases;—not to general rules capable of direct verification; but to notions vague, distant, and vast, which we cannot bring into contact with facts, because they belong to a different region from the facts; as when we connect natural events with moral or historical causes, or seek spiritual meanings in the properties of number and figure. Thus the character of Mysticism is, that it refers particulars, not to generalizations homogeneous and immediate, but to such as are heterogeneous and remote; to which we must add, that the process of this reference is not a calm act of the intellect, but is accompanied with a glow of enthusiastic feeling.

CHAPTER II.

INDUCTION OF COPERNICUS.—THE HELIOCENTRIC THEORY ASSERTED ON FORMAL GROUNDS.

IT will be recollected that the *formal* are opposed to the *physical* grounds of a theory; the former term indicating that it gives a satisfactory account of the relations of the phenomena in Space and Time, that is, of the Motions themselves; while the latter expression implies further that we include in our explanation the Causes of the motions, the laws of Force and Matter. The strongest of the considerations by which Copernicus was led to invent and adopt his system of the universe were of the former kind. He was dissatisfied, he says, in his Preface addressed to the Pope, with the want of symmetry in the Eccentric Theory, as it prevailed in his days; and weary of the uncertainty of the mathematical traditions. He then sought through all the works of philosophers, whether any had held opinions concerning the motions of the world, different from those received in the established mathematical schools. He found, in ancient authors, accounts of Philolaus and others, who had asserted the motion of the earth. 'Then,' he adds, 'I, too, began to meditate concerning the motion of the earth: and though it appeared an absurd opinion, yet since I knew that, in previous times, others had been allowed the privilege of feigning what circles they chose, in order to explain the phenomena, I conceived that I also might take the liberty of trying whether, on the supposition of the earth's motion, it was possible to find better explanations than the ancient ones, of the revolutions of the celestial orbs.

'Having then assumed the motions of the earth, which are hereafter explained, by laborious and long observation I at length found, that if the motions of the other planets be compared with the revolution of the earth, not only their phenomena follow from the suppositions, but also that the several orbs, and the whole system, are so connected in order and magnitude, that no one part can be transposed without disturbing the rest, and introducing confusion into the whole universe.'

Thus the satisfactory explanation of the apparent motions of the planets, and the simplicity and symmetry of the system, were the grounds on which Copernicus adopted his theory; as the craving for these qualities was the feeling which led him to seek for a new theory. It is manifest that in this, as in other cases of discovery, a clear and steady possession of abstract Ideas, and an aptitude in comprehending real Facts under these general conceptions, must have been leading characters in the discoverer's mind. He must have had a good geometrical head, and great astronomical knowledge. He must have seen, with peculiar distinctness, the consequences which flowed from his suppositions as to the relations of space and time,—the apparent motions which resulted from the assumed real ones; and he must also have known well all the irregularities of the apparent motions for which he had to account. We find indications of these qualities in his expressions. A steady and calm contemplation of the theory is what he asks for, as the main requisite to its reception. If you suppose the earth to revolve and the heaven to be at rest, you will find, he says, '*si serio animadvertas,*' if you think steadily, that the apparent diurnal motion will follow. And after alleging his reasons for his system, he says,[1] 'We are, therefore, not ashamed to confess, that the whole of the space within the orbit of the moon, along with the center of the earth, moves round the sun in a year among the other planets; the magnitude of the world being so great, that the distance of the earth from the sun has no apparent magnitude when compared with the sphere of the fixed stars.' 'All which things, though they be difficult and almost inconceivable, and against the opinion of the majority, yet, in the sequel, by God's favour, we will make clearer

---

[1] Nicolai Copernici Torinensis *de Revolutionibus Orbium Cœlestium Libri VI.* Norimbergæ, M.D.XLIII. p. 9.

than the sun, at least to those who are not ignorant of mathematics.'

It will easily be understood, that since the ancient geocentric hypothesis ascribed to the planets those motions which were apparent only, and which really arose from the motion of the earth round the sun in the new hypothesis, the latter scheme must much simplify the planetary theory. Kepler[2] enumerates eleven motions of the Ptolemaic system, which are at once exterminated and rendered unnecessary by the new system. Still, as the real motions, both of the earth and the planets, are unequable, it was requisite to have some mode of representing their inequalities; and, accordingly, the ancient theory of eccentrics and epicycles was retained, so far as was requisite for this purpose. The planets revolved round the sun by means of a Deferent, and a great and small Epicycle; or else by means of an Eccentric and Epicycle, modified from Ptolemy's, for reasons which we shall shortly mention. This mode of representing the motions of the planets continued in use, till it was expelled by the discoveries of Kepler.

Besides the daily rotation of the earth on its axis, and its annual circuit about the sun, Copernicus attributed to the axis a 'motion of declination,' by which, during the whole annual revolution, the pole was constantly directed towards the same part of the heavens. This constancy in the absolute direction of the axis, or its moving parallel to itself, may be more correctly viewed as not indicating any separate motion. The axis continues in the same direction, because there is nothing to make it change its direction; just as a straw, lying on the surface of a cup of water, continues to point nearly in the same direction when the cup is carried round a room. And this was noticed by Copernicus's adherent, Rothmah,[3] a few years after the publication of the work *De Revolutionibus*. 'There is no occasion,' he says, in a letter to Tycho Brahe, 'for the

---

[2] *Myst. Cosm.* cap. 1.
[3] Tycho. Epist. i. p. 184. A.D. 1590.

triple motion of the earth: the annual and diurnal motions suffice.' This errour of Copernicus, if it be looked upon as an errour, arose from his referring the position of the axis to a limited space, which he conceived to be carried round the sun along with the earth, instead of referring it to fixed or absolute space. When, in a Planetarium, (a machine in which the motions of the planets are imitated,) the earth is carried round the sun by being fastened to a material radius, it is requisite to give a motion to the axis by *additional* machinery, in order to enable it to *preserve* its parallelism. A similar confusion of geometrical conception, produced by a double reference to absolute space and to the center of revolution, often leads persons to dispute whether the moon, which revolves about the earth, always turning to it the same face, revolves about her axis or not.

It is also to be noticed that the precession of the equinoxes made it necessary to suppose the axis of the earth to be not *exactly* parallel to itself, but to deviate from that position by a slight annual difference. Copernicus erroneously supposes the precession to be unequable; and his method of explaining this change, which is simpler than that of the ancients, becomes more simple still, when applied to the true state of the facts.

The tendencies of our speculative nature, which carry us onwards in pursuit of symmetry and rule, and which thus produced the theory of Copernicus, as they produce all theories, perpetually show their vigour by overshooting their mark. They obtain something by aiming at much more. They detect the order and connexion which exist, by imagining relations of order and connexion which have no existence. Real discoveries are thus mixed with baseless assumptions; profound sagacity is combined with fanciful conjecture; not rarely, or in peculiar instances, but commonly, and in most cases; probably in all, if we could read the thoughts of the discoverers as we read the books of Kepler. To try wrong guesses is apparently the only way to hit upon right ones. The character of the true

philosopher is, not that he never conjectures hazard-
ously, but that his conjectures are clearly conceived
and brought into rigid contact with facts. He sees
and compares distinctly the ideas and the things,—the
relations of his notions to each other and to phenomena.
Under these conditions it is not only excusable, but
necessary for him, to snatch at every semblance of
general rule;—to try all promising forms of simplicity
and symmetry.

Copernicus is not exempt from giving us, in his work,
an example of this character of the inventive spirit.
The axiom that the celestial motions must be *circular*
and *uniform*, appeared to him to have strong claims to
acceptation; and his theory of the inequalities of the
planetary motions is fashioned upon it. His great
desire was to apply it more rigidly than Ptolemy had
done. The time did not come for rejecting this axiom,
till the observations of Tycho Brahe and the calculations of Kepler had been made.

I shall not attempt to explain, in detail, Copernicus's
system of the planetary inequalities. He retained
epicycles and eccentrics, altering their centers of
motion; that is, he retained what was *true* in the old
system, *translating* it into his own. The peculiarities
of his method consisted in making such a combination
of epicycles as to supply the place of the *equant*,[4] and
to make all the motions equable about the centers of
motion. This device was admired for a time, till
Kepler's elliptic theory expelled it, with all other forms
of the theory of epicycles: but we must observe that
Copernicus was aware of some of the discrepancies
which belonged to that theory as it had, up to that
time, been propounded. In the case of Mercury's
orbit, which is more eccentric than that of the other
planets, he makes suppositions which are complex
indeed, but which show his perception of the imperfection of the common theory; and he proposes a new
theory of the moon, for the very reason which did at
last overturn the doctrine of epicycles, namely, that the

---

[4] See B. iii. Chap. iii. Sect. 7.

ratio of their distances from the earth at different times was inconsistent with the circular hypothesis.[5]

It is obvious, that, along with his mathematical clearness of view, and his astronomical knowledge, Copernicus must have had great intellectual boldness and vigour, to conceive and fully develope a theory so different as his was, from all received doctrines. His pupil and expositor, Rheticus, says to Schener, ' I beg you to have this opinion concerning that learned man, my Preceptor; that he was an ardent admirer and follower of Ptolemy; but when he was compelled by phenomena and demonstration, he thought he did well to aim at the same mark at which Ptolemy had aimed, though with a bow and shafts of a very different material from his. We must recollect what Ptolemy says, Δεῖ δ' ἐλευθέρον εἶναι τῇ γνώμῃ τὸν μέλλοντα φιλοσοφεῖν. ' He who is to follow philosophy must be a freeman in mind.' Rheticus then goes on to defend his master from the charge of disrespect to the ancients : ' That temper,' he says, ' is alien from the disposition of every good man, and most especially from the spirit of philosophy, and from no one more utterly than from my Preceptor. He was very far from rashly rejecting the opinions of ancient philosophers, except for weighty reasons and irresistible facts, through any love of novelty. His years, his gravity of character, his excellent learning, his magnanimity and nobleness of spirit, are very far from having any liability to such a temper, which belongs either to youth, or to ardent and light minds, or to those τῶν μέγα φρονούντων ἐπὶ θεωρίᾳ μικρῇ, ' who think much of themselves and know little,' as Aristotle says.' Undoubtedly this deference for the great men of the past, joined with the talent of seizing the spirit of their methods when the letter of their theories is no longer tenable, *is* the true mental constitution of discoverers.

Besides the intellectual energy which was requisite in order to construct a system of doctrines so novel as those of Copernicus, some courage was necessary to the

---

[5] *De Rev.* iv. c. 2.

publication of such opinions; certain, as they were, to be met, to a great extent, by rejection and dispute, and perhaps by charges of heresy and mischievous tendency. This last danger, however, must not be judged so great as we might infer from the angry controversies and acts of authority which occurred in Galileo's time. The Dogmatism of the stationary period, which identified the cause of philosophical and religious truth, had not yet distinctly felt itself attacked by the advance of physical knowledge; and therefore had not begun to look with alarm on such movements. Still, the claims of Scripture and of ecclesiastical authority were asserted as paramount on all subjects; and it was obvious that many persons would be disquieted or offended with the new interpretation of many scriptural expressions, which the true theory would make necessary. This evil Copernicus appears to have foreseen; and this and other causes long withheld him from publication. He was himself an ecclesiastic; and, by the patronage of his maternal uncle, was prebendary of the church of St. John at Thorn, and a canon of the church of Frauenburg, in the diocese of Ermeland.[6] He had been a student at Bologna, and had taught mathematics at Rome in the year 1500; and he afterwards pursued his studies and observations at his residence near the mouth of the Vistula.[7] His discovery of his system must have occurred before 1507, for in 1543 he informs Pope Paulus the Third, in his dedication, that he had kept his book by him for four times the nine years recommended by Horace, and then only published it at the earnest entreaty of his friend Cardinal Schomberg, whose letter is prefixed to the work. 'Though I know,' he says, 'that the thoughts of a philosopher do not depend on the judgment of the many, his study being to seek out truth in all things as far as that is permitted by God to human reason: yet when I considered,' he adds, 'how absurd my doctrine would appear, I long hesitated whether I should publish my

---

[6] Rheticus, *Nar.* p 94.　　　　[7] Riccioli.

book, or whether it were not better to follow the example of the Pythagoreans and others, who delivered their doctrines only by tradition and to friends.' It will be observed that he speaks here of the opposition of the established school of Astronomers, not of Divines. The latter, indeed, he appears to consider as a less formidable danger. ' If perchance,' he says at the end of his preface, 'there be ματαιολόγοι, vain babblers, who knowing nothing of mathematics, yet assume the right of judging on account of some place of Scripture perversely wrested to their purpose, and who blame and attack my undertaking; I heed them not, and look upon their judgments as rash and contemptible.' He then goes on to show that the globular figure of the earth (which was, of course, at that time, an undisputed point among astronomers,) had been opposed on similar grounds by Lactantius, who, though a writer of credit in other respects, had spoken very childishly in that matter. In another epistle prefixed to the work (by Andreas Osiander), the reader is reminded that the hypotheses of astronomers are not necessarily asserted to be true, by those who propose them, but only to be a way of *representing* facts. We may observe that, in the time of Copernicus, when the motion of the earth had not been connected with the physical laws of matter and motion, it could not be considered so distinctly real as it necessarily was held to be in after times.

The delay of the publication of Copernicus's work brought it to the end of his life: he died in the year 1543, in which it was published. It was entitled *De Revolutionibus Orbium Cœlestium Libri VI.* He received the only copy he ever saw on the day of his death, and never opened it: he had then, says Gassendi, his biographer, other cares. His system was, however, to a certain extent, promulgated, and his fame diffused before that time. Cardinal Schomberg, in his letter of 1536, which has been already mentioned, says, 'Some years ago, when I heard tidings of your merit by the constant report of all persons, my affection for you was augmented, and I congratulated the men of our time, among whom you flourish in so much honour. For I had understood that you were not only acquainted with the discoveries of ancient mathematicians, but also had formed a new system of the world, in which you teach that the Earth moves, the Sun occupies the lowest, and consequently, the middle place, the sphere of the fixed stars remains immoveable and fixed, and the Moon, along with the elements included in her sphere, placed between the orbits (*cœlum*) of Mars and Venus, travels round the sun in a yearly revolution.'[8] The writer goes on to say that he has heard that Copernicus has written a book (*Commentarios*), in which this system is applied to the construction of Tables of the Planetary Motions (*erraticarum stellarum*). He then proceeds to entreat him earnestly to publish his lucubrations.

This letter is dated 1536, and implies that the work of Copernicus was then written, and known to persons who studied astronomy. Delambre says that Achilles Gassarus of Lindau, in a letter dated 1540, sends to his friend George Vogelin of Constance, the book *De Revolutionibus*. But Mr. De Morgan[9] has pointed out that the printed work which Gassarus sent to Vogelin was the *Narratio* by Rheticus of Feldkirch, an eulogium

---

[8] This passage has so important a place in the history, that I will give it in the original :—' Intellexeram te non modo veterum mathematicorum inventa egregie callere sed etiam novam mundi rationem constituisse: Qua doceas terram moveri : solem imum mundi, atque medium locum obtinere : cœlum octavum immotum atque fixum perpetuo manere : Lunam se una cum inclusis suæ spheræ elementis, inter Martis et Veneris cœlum sitam, anniversario cursu circum solem convertere. Atque de hac tota astronomiæ ratione commentarios a te confectos esse, ac erraticarum stellarum motus calculis subductos tabulis te contulisse, maxima omnium cum admiratione. Quamobrem vir doctissime, nisi tibi molestus sum, te etiam atque etiam oro vehementer ut hoc tuum inventum studiosis communices, et tuas de mundi sphæra lucubrationes, una cum Tabulis et si quid habes præterea quod ad eandem rem pertineat primo quoque tempore ad me mittas.'

[9] *Ast. Mod.* i. p. 138. I owe this and many other corrections to the personal kindness of Mr. De Morgan.

of Copernicus and his system prefixed to the second edition of the *De Revolutionibus*, which appeared in 1566. In this Narration, Rheticus speaks of the work of Copernicus as a Palingenesia, or New Birth of astronomy. Rheticus, it appears, had gone to Copernicus for the purpose of getting knowledge about triangles and trigonometrical tables, and had had his attention called to the heliocentric theory, of which he became an ardent admirer. He speaks of his ' Preceptor' with strong admiration, as we have seen. ' He appears to me,' says he, ' more to resemble Ptolemy than any other astronomer.' This, it must be recollected, was selecting the highest known subject of comparison.

CHAPTER II.

IN order that we may the more clearly consider the bearing of this, the greatest scientific discovery ever made, we shall resolve it into the partial propositions of which it consists. Of these we may enumerate five. The doctrine of universal gravitation asserts,

1. That the force by which the *different* planets are attracted to the sun is in the inverse proportion of the squares of their distances;

2. That the force by which the *same* planet is attracted to the sun, in different parts of its orbit, is also in the inverse proportion of the squares of the distances;

3. That the *earth* also exerts such a force on the *moon*, and that this force is identical with the force of *gravity*;

4. That bodies thus act on *other* bodies, besides those which revolve round them; thus, that the sun exerts such a force on the moon and satellites, and that the planets exert such forces on *one another;*

5. That this force, thus exerted by the general masses of the sun, earth, and planets, arises from the attraction *of each particle* of these masses; which attraction follows the above law, and belongs to all matter alike.

The history of the establishment of these five truths will be given in order.

1. *Sun's Force on Different Planets.*—With regard to the first of the above five propositions, that the different planets are attracted to the sun by a force which is inversely as the square of the distance,

Newton had so far been anticipated, that several persons had discovered it to be true, or nearly true; that is, they had discovered that if the orbits of the planets were circles, the proportions of the central force to the inverse square of the distance would follow from Kepler's third law, of the sesquiplicate proportion of the periodic times. As we have seen, Huyghens' theorems would have proved this, if they had been so applied; Wren knew it; Hooke not only knew it, but claimed a prior knowledge to Newton; and Halley had satisfied himself that it was at least nearly true, before he visited Newton. Hooke was reported to Newton at Cambridge, as having applied to the Royal Society to do him justice with regard to his claims; but when Halley wrote and informed Newton (in a letter dated June 29, 1686), that Hooke's conduct 'had been represented in worse colours than it ought,' Newton inserted in his book a notice of these his predecessors, in order, as he said, ' to compose the dispute.'[1] This notice appears in a Scholium to the fourth Proposition of the *Principia*, which states the general law of revolutions in circles. ' The case of the sixth corollary,' Newton there says, ' obtains in the celestial bodies, as has been separately inferred by our countrymen, Wren, Hooke, and Halley;' he soon after names Huyghens, ' who, in his excellent treatise *De Horologio Oscillatorio*, compares the force of gravity with the centrifugal forces of revolving bodies.'

The two steps requisite for this discovery were, to propose the motions of the planets as simply a mechanical problem, and to apply mathematical reasoning so as to solve this problem, with reference to Kepler's third law considered as a fact. The former step was a consequence of the mechanical discoveries of Galileo and his school; the result of the firm and clear place which these gradually obtained in men's minds, and of the utter abolition of all the notions of solid spheres by Kepler. The mathematical step required no small mathematical powers; as appears, when we consider

---

[1] *Biog. Brit.* folio, art. *Hooke.*

that this was the first example of such a problem, and
that the method of limits, under all its forms, was at
this time in its infancy, or rather, at its birth. Accordingly, even this step, though much the easiest in the
path of deduction, no one before Newton completely
executed.

2. *Force in different Points of an Orbit.*—The inference of the law of the force from Kepler's two laws
concerning the elliptical motion, was a problem quite
different from the preceding, and much more difficult;
but the dispute with respect to priority in the two
propositions was intermingled. Borelli, in 1666, had,
as we have seen, endeavoured to reconcile the general
form of the orbit with the notion of a central attractive force, by taking centrifugal force into the account;
and Hooke, in 1679, had asserted that the result of
the law of the inverse square in the force of the earth
would be an ellipse,[2] or a curve like an ellipse.[3] But
it does not appear that this was anything more than
a conjecture. Halley says[4] that 'Hooke, in 1683,
told him he had demonstrated all the laws of the
celestial motions by the reciprocally duplicate proportion of the force of gravity; but that, being offered
forty shillings by Sir Christopher Wren to produce
such a demonstration, his answer was, that he had it,
but would conceal it for some time, that others, trying
and failing, might know how to value it when he
should make it public.' Halley, however, truly observes, that after the publication of the demonstration
in the *Principia*, this reason no longer held; and adds,
' I have plainly told him, that unless he produce
another differing demonstration, and let the world
judge of it, neither I nor any one else can believe it.'

Newton allows that Hooke's assertions in 1679 gave
occasion to his investigation on this point of the theory.
His demonstration is contained in the second and
third Sections of the *Principia*. He first treats of the

[2] Newton's Letter, *Biog. Brit.*, Hooke, p. 2660.
[3] Birch's *Hist. R. S.*, Wallis's Life.
[4] *Enc. Brit.*, Hooke, p. 2660.

general law of central forces in any curve; and then, on account, as he states, of the application to the motion of the heavenly bodies, he treats of the case of force varying inversely as the square of the distance, in a more diffuse manner.

In this, as in the former portion of his discovery, the two steps were, the proposing the heavenly motions as a mechanical problem, and the solving this problem. Borelli and Hooke had certainly made the former step, with considerable distinctness; but the mathematical solution required no common inventive power.

Newton seems to have been much ruffled by Hooke's speaking slightly of the value of this second step; and is moved in return to deny Hooke's pretensions with some asperity, and to assert his own. He says, in a letter to Halley, ' Borelli did something in it, and wrote modestly; he (Hooke) has done nothing; and yet written in such a way as if he knew and had sufficiently hinted all but what remained to be determined by the drudgery of calculations and observations; excusing himself from that labour by reason of his other business; whereas he should rather have excused himself by reason of his inability: for it is very plain, by his words, he knew not how to go about it. Now is not this very fine? Mathematicians that find out, settle, and do all the business, must content themselves with being nothing but dry calculators and drudges; and another that does nothing but pretend and grasp at all things, must carry away all the inventions, as well of those that were to follow him as of those that went before.' This was written, however, under the influence of some degree of mistake; and in a subsequent letter, Newton says, ' Now I understand he was in some respects misrepresented to me, I wish I had spared the postscript to my last,' in which is the passage just quoted. We see, by the melting away of rival claims, the undivided honour which belongs to Newton, as the real discoverer of the proposition now under notice. We may add, that in the sequel of the third Section of the *Principia*, he has traced its consequences, and solved various problems flowing from it with his usual fertility and beauty of mathematical resource; and has there shown the necessary connexion of Kepler's third law with his first and second.

3. *Moon's Gravity to the Earth.*—Though others had considered cosmical forces as governed by the general laws of motion, it does not appear that they had identified such forces with the force of terrestrial gravity. This step in Newton's discoveries has generally been the most spoken of by superficial thinkers; and a false kind of interest has been attached to it, from the story of its being suggested by the fall of an apple. The popular mind is caught by the character of an eventful narrative which the anecdote gives to this occurrence; and by the antithesis which makes a profound theory appear the result of a trivial accident. How inappropriate is such a view of the matter we shall soon see. The narrative of the progress of Newton's thoughts, is given by Pemberton (who had it from Newton himself) in his preface to his *View of Newton's Philosophy*, and by Voltaire, who had it from Mrs. Conduit, Newton's niece.[5] 'The first thoughts,' we are told, 'which gave rise to his *Principia*, he had when he retired from Cambridge, in 1666, on account of the plague, (he was then twenty-four years of age). As he sat alone in a garden, he fell into a speculation on the power of gravity; that as this power is not found sensibly diminished at the remotest distance from the center of the earth to which we can rise, neither at the tops of the loftiest buildings, nor even on the summits of the highest mountains, it appeared to him reasonable to conclude that this power must extend much further than was usually thought: Why not as high as the moon? said he to himself; and if so, her motion must be influenced by it; perhaps she is retained in her orbit thereby.'

The thought of cosmical gravitation was thus distinctly brought into being: and Newton's superiority here was, that he conceived the celestial motions as distinctly as the motions which took place close to

---

[5] *Elémens de Phil. de Newton*, 3me partie, chap. iii.

him;—considered them as of the same kind, and applied
the same rules to each, without hesitation or obscurity.
But so far, this thought was merely a guess: its occurrence showed the activity of the thinker; but to give
it any value, it required much more than a ' why not?'
—a ' perhaps.' Accordingly, Newton's ' why not?'
was immediately succeeded by his ' if so, what then?'
His reasoning was, that if gravity reach to the moon,
it is probably of the same kind as the central force of
the sun, and follows the same rule with respect to the
distance. What is this rule? We have already seen
that, by calculating from Kepler's laws, and supposing
the orbits to be circles, the rule of the force appears to
be the inverse duplicate proportion of the distance;
and this, which had been current as a conjecture among
the previous generation of mathematicians, Newton
had already proved by indisputable reasonings, and
was thus prepared to proceed in his train of inquiry.
If, then, he went on, pursuing his train of thought,
the earth's gravity extend to the moon, diminishing
according to the inverse square of the distance, will
it, at the moon's orbit, be of the proper magnitude for
retaining her in her path? Here again came in calculation, and a calculation of extreme interest; for how
important and how critical was the decision which
depended on the resulting numbers? According to
Newton's calculations, made at this time, the moon, by
her motion in her orbit, was deflected from the tangent
every minute through a space of thirteen feet. But
by noticing the space through which bodies would fall
in one minute at the earth's surface, and supposing
this to be diminished in the ratio of the inverse square,
it appeared that gravity would, at the moon's orbit,
draw a body through more than fifteen feet. The difference seems small, the approximation encouraging,
the theory plausible; a man in love with his own
fancies would readily have discovered or invented some
probable cause of this difference. But Newton ac-
quiesced in it as a disproof of his conjecture, and ' laid
aside at that time any further thoughts of this matter;'
thus resigning a favourite hypothesis, with a candour

and openness to conviction not inferior to Kepler, though his notion had been taken up on far stronger and sounder grounds than Kepler dealt in; and without even, so far as we know, Kepler's regrets and struggles. Nor was this levity or indifference; the idea, though thus laid aside, was not finally condemned and abandoned. When Hooke, in 1679, contradicted Newton on the subject of the curve described by a falling body, and asserted it to be an ellipse, Newton was led to investigate the subject, and was then again conducted, by another road, to the same law of the inverse square of the distance. This naturally turned his thoughts to his former speculations. Was there really no way of explaining the discrepancy which this law gave, when he attempted to reduce the moon's motion to the action of gravity? A scientific operation then recently completed, gave the explanation at once. He had been mistaken in the magnitude of the earth, and consequently in the distance of the moon, which is determined by measurements of which the earth's radius is the base. He had taken the common estimate, current among geographers and seamen, that sixty English miles are contained in one degree of latitude. But Picard, in 1670, had measured the length of a certain portion of the meridian in France, with far greater accuracy than had yet been attained; and this measure enabled Newton to repeat his calculations with these amended data. We may imagine the strong curiosity which he must have felt as to the result of these calculations. His former conjecture was now found to agree with the phenomena to a remarkable degree of precision. This conclusion, thus coming after long doubts and delays, and falling in with the other results of mechanical calculation for the solar system, gave a stamp from that moment to his opinions, and through him to those of the whole philosophical world.

[2nd Ed.] [Dr. Robison (*Mechanical Philosophy*, p. 288) says that Newton having become a member of the Royal Society, there learned the accurate measurement of the earth by Picard, differing very much from the estimation by which he had made his calculations in 1666. And M. Biot, in his life of Newton, published in the *Biographie Universelle*, says, '*According to conjecture*, about the month of June, 1682, Newton being in London, at a meeting of the Royal Society, mention was made of the new measure of a degree of the earth's surface, recently executed in France by Picard; and great praise was given to the care which had been employed in making this measure exact.'

I had adopted this conjecture as a fact in my first edition; but it has been pointed out by Prof. Rigaud (*Historical Essay on the First Publication of the Principia*, 1838), that Picard's measurement was probably well known to the Fellows of the Royal Society as early as 1675, there being an account of the results of it given in the *Philosophical Transactions* for that year. Newton appears to have discovered the method of determining that a body might describe an ellipse when acted upon by a force residing in the focus, and varying inversely as the square of the distance, in 1679, upon occasion of his correspondence with Hooke. In 1684, at Halley's request, he returned to the subject, and in February, 1685, there was inserted in the Register of the Royal Society a paper of Newton's (*Isaaci Newtoni Propositiones de Motu*) which contained some of the principal Propositions of the first two Books of the *Principia*. This paper, however, does not contain the Proposition 'Lunam gravitare in terram,' nor any of the other propositions of the third Book. The *Principia* was printed in 1686 and 7, apparently at the expense of Halley. On the 6th of April, 1687, the Third Book was presented to the Royal Society.]

It does not appear, I think, that before Newton, philosophers in general had supposed that terrestrial gravity was the very force by which the moon's motions are produced. Men had, as we have seen, taken up the conception of such forces, and had probably called them gravity: but this was done only to explain, by analogy, what *kind* of forces they were, just as at other times they compared them with magnetism; and it did not imply that terrestrial gravity was a force which acted in the celestial spaces. After Newton had discovered that this was so, the application of the term 'gravity' did undoubtedly convey such a suggestion; but we should err if we inferred from this coincidence of expression that the notion was commonly entertained before him. Thus Huyghens appears to use language which may be mistaken, when he says,[6] that Borelli was of opinion that the primary planets were urged by 'gravity' towards the sun, and the satellites towards the primaries. The notion of terrestrial gravity, as being actually a cosmical force, is foreign to all Borelli's speculations.[7] But Horrox, as early as 1635, appears to have entertained the true view on this subject, although vitiated by Keplerian errours concerning the connexion between the rotation of the central body and its effect on the body which revolves about it. Thus he says,[8] that the emanation of the earth carries a projected stone along with the motion of the earth, just in the same way as it carries the moon in her orbit; and that this force is greater on the stone than on the moon, because the distance is less.

The Proposition in which Newton has stated the discovery of which we are now speaking, is the fourth of his third Book : 'That the moon gravitates to the earth, and by the force of gravity is perpetually deflected from a rectilinear motion, and retained in her orbit.' The proof consists in the numerical calculation, of which he only gives the elements, and points out the method; but we may observe, that no small degree of knowledge of the way in which astronomers had obtained these elements, and judgment in selecting among them, were necessary : thus, the mean distance of the moon had been made as little as fifty-six and a half semi-diameters of the earth by Tycho, and as

---

[6] *Cosmotheros*, l. 2. p. 720.

[7] I have found no instance in which the word is so used by him.

[8] *Astronomia Kepleriana defensa et promota*, cap. 2. See further on this subject in the *Additions* to this volume.

much as sixty-two and a half by Kircher: Newton
gives good reasons for adopting sixty-one.

The term 'gravity,' and the expression 'to gravitate,' which, as we have just seen, Newton uses of the
moon, were to receive a still wider application in consequence of his discoveries; but in order to make this
extension clearer, we consider it as a separate step.

4. *Mutual Attraction of all the Celestial Bodies.*—
If the preceding parts of the discovery of gravitation
were comparatively easy to conjecture, and difficult to
prove; this was much more the case with the part of
which we have now to speak, the attraction of other
bodies, besides the central ones, upon the planets and
satellites. If the mathematical calculation of the
unmixed effect of a central force required transcendent
talents, how much must the difficulty be increased,
when other influences prevented those first results
from being accurately verified, while the deviations
from accuracy were far more complex than the original
action! If it had not been that these deviations,
though surprisingly numerous and complicated in their
nature, were very small in their quantity, it would
have been impossible for the intellect of man to deal
with the subject; as it was, the struggle with its difficulties is even now a matter of wonder.

The conjecture that there is some mutual action of
the planets, had been put forth by Hooke in his
Attempt to prove the Motion of the Earth, (1674.) It
followed, he said, from his doctrine, that not only the
sun and the moon act upon the course and motion of
the earth, but that Mercury, Venus, Mars, Jupiter,
and Saturn, have also, by their attractive power, a considerable influence upon the motion of the earth, and
the earth in like manner powerfully affects the motions
of those bodies. And Borelli, in attempting to form
'theories' of the satellites of Jupiter, had seen, though
dimly and confusedly, the probability that the sun
would disturb the motions of these bodies. Thus he
says, (cap. 14,) ' How can we believe that the Medicean
globes are not, like other planets, impelled with a
greater velocity when they approach the sun: and

thus they are acted upon by two moving forces, one of which produces their proper revolution about Jupiter, the other regulates their motion round the sun.' And in another place, (cap. 20,) he attempts to show an effect of this principle upon the inclination of the orbit; though, as might be expected, without any real result.

The case which most obviously suggests the notion that the sun exerts a power to disturb the motions of secondary planets about primary ones, might seem to be our own moon; for the great inequalities which had hitherto been discovered, had all, except the first, or elliptical anomaly, a reference to the position of the sun. Nevertheless, I do not know that any one had attempted thus to explain the curiously irregular course of the earth's attendant. To calculate, from the disturbing agency, the amount of the irregularities, was a problem which could not, at any former period, have been dreamt of as likely to be at any time within the verge of human power.

Newton both made the step of inferring that there were such forces, and, to a very great extent, calculated the effects of them. The inference is made on mechanical principles, in the sixth Theorem of the third Book of the *Principia;*—that the moon is attracted by the sun, as the earth is;—that the satellites of Jupiter and Saturn are attracted as the primaries are; in the same manner, and with the same forces. If this were not so, it is shown that these attendant bodies could not accompany the principal ones in the regular manner in which they do. All those bodies at equal distances from the sun would be equally attracted.

But the complexity which must occur in tracing the results of this principle will easily be seen. The satellite and the primary, though nearly at the same distance, and in the same direction, from the sun, are not exactly so. Moreover the difference of the distances and of the directions is perpetually changing; and if the motion of the satellite be elliptical, the cycle of change is long and intricate: on this account alone the effects of the sun's action will inevitably follow cycles as long and as perplexed as those of the positions. But on another account they will be still more complicated; for in the continued action of a force, the effect which takes place at first, modifies and alters the effect afterwards. The result at any moment is the sum of the results in preceding instants: and since the terms, in this series of instantaneous effects, follow very complex rules, the sums of such series will be, it might be expected, utterly incapable of being reduced to any manageable degree of simplicity.

It certainly does not appear that any one but Newton could make any impression on this problem, or course of problems. No one for sixty years after the publication of the *Principia*, and, with Newton's methods, no one up to the present day, had added anything of any value to his deductions. We know that he calculated all the principal lunar inequalities; in many of the cases, he has given us his processes; in others, only his results. But who has presented, in his beautiful geometry, or deduced from his simple principles, any of the inequalities which he left untouched? The ponderous instrument of synthesis, so effective in his hands, has never since been grasped by one who could use it for such purposes; and we gaze at it with admiring curiosity, as on some gigantic implement of war, which stands idle among the memorials of ancient days, and makes us wonder what manner of man he was who could wield as a weapon what we can hardly lift as a burden.

It is not necessary to point out in detail the sagacity and skill which mark this part of the *Principia*. The mode in which the author obtains the effect of a disturbing force in producing a motion of the apse of an elliptical orbit (the ninth Section of the first Book), has always been admired for its ingenuity and elegance. The general statement of the nature of the principal inequalities produced by the sun in the motion of a satellite, given in the sixty-sixth Proposition, is, even yet, one of the best explanations of such action; and the calculations of the quantity of the effects in the third Book, for instance, the *variation* of the moon, the *motion of the nodes* and its inequalities, the *change of inclination* of the orbit,—are full of beautiful and efficacious artifices. But Newton's inventive faculty was exercised to an extent greater than these published investigations show. In several cases he has suppressed the demonstration of his method, and given us the result only; either from haste, or from mere weariness, which might well overtake one who, while he was struggling with facts and numbers, with difficulties of conception and practice, was aiming also at that geometrical elegance of exposition, which he considered as alone fit for the public eye. Thus, in stating the effect of the eccentricity of the moon's orbit upon the motion of the apogee, he says,[9] 'The computations, as too intricate and embarrassed with approximations, I do not choose to introduce.'

The computations of the theoretical motion of the moon being thus difficult, and its irregularities numerous and complex, we may ask, whether Newton's reasoning was sufficient to establish this part of his theory; namely, that her actual motions arise from her gravitation to the sun. And to this we may reply, that it was sufficient for that purpose,—since it showed that, from Newton's hypothesis, inequalities must result, following the laws which the moon's inequalities were known to follow;—since the amount of the inequalities given by the theory agreed nearly with the rules which astronomers had collected from observation;—and since, by the very intricacy of the calculation, it was rendered probable, that the first results might be somewhat inaccurate, and thus might give rise to the still remaining differences between the calculations and the facts. A *Progression of the Apogee;* a *Regression of the Nodes;* and, besides the Elliptical, or first Inequality, an inequality, following the law of the *Evection*, or second inequality discovered by Ptolemy; another, following the law of the *Variation* discovered by Tycho;—were pointed out in the first

---

[9] Schol. to Prop. 35, first edit.

edition of the *Principia*, as the consequences of the theory. Moreover, the quantities of these inequalities were calculated and compared with observation with the utmost confidence, and the agreement in most instances was striking. The Variation agreed with Halley's recent observations within a minute of a degree.[10] The Mean Motion of the Nodes in a year agreed within less than one-hundredth of the whole.[11] The Equation of the Motion of the Nodes also agreed well.[12] The Inclination of the Plane of the Orbit to the ecliptic, and its changes, according to the different situations of the nodes, likewise agreed.[13] The Evection has been already noticed as encumbered with peculiar difficulties: here the accordance was less close. The Difference of the daily progress of the Apogee in syzygy, and its daily Regress in Quadratures, is, Newton says, '$4\frac{1}{4}$ minutes by the Tables, $6\frac{2}{3}$ by our calculation.' He boldly adds, ' I suspect this difference to be due to the fault of the Tables.' In the second edition (1711) he added the calculation of several other inequalities, as the *Annual Equation*, also discovered by Tycho; and he compared them with more recent observations made by Flamsteed at Greenwich; but even in what has already been stated, it must be allowed that there is a wonderful accordance of theory with phenomena, both being very complex in the rules which they educe.

The same theory which gave these Inequalities in the motion of the Moon produced by the disturbing force of the sun, gave also corresponding Inequalities in the motions of the Satellites of other planets, arising from the same cause; and likewise pointed out the necessary existence of irregularities in the motions of the Planets arising from their mutual attraction. Newton gave propositions by which the Irregularities of the motion of Jupiter's moons might be deduced from those of our own;[14] and it was shown that the motions of their nodes would be slow by theory, as

---

10 B. iii. Prop. 29. 11 Prop. 32. 12 Prop. 33.
13 Prop. 35. 14 B. i. Prop. 66.

Flamsteed had found it to be by observation.[15]   But
Newton did not attempt to calculate the effect of the
mutual action of the planets, though he observes, that
in the case of Jupiter and Saturn this effect is too
considerable to be neglected;[16] and he notices in the
second edition,[17] that it follows from the theory of
gravity, that the aphelia of Mercury, Venus, the
Earth, and Mars, slightly progress.

In one celebrated instance, indeed, the deviation of
the theory of the *Principia* from observation was
wider, and more difficult to explain; and as this deviation for a time resisted the analysis of Euler and
Clairaut, as it had resisted the synthesis of Newton, it
at one period staggered the faith of mathematicians in
the exactness of the law of the inverse square of the
distance.   I speak of the Motion of the Moon's Apogee, a problem which has already been referred to;
and in which Newton's method, and all the methods
which could be devised for some time afterwards, gave
only half the observed motion; a circumstance which
arose, as was discovered by Clairaut in 1750, from the
insufficiency of the method of approximation.   Newton does not attempt to conceal this discrepancy.
After calculating what the motion of apse would be,
upon the assumption of a disturbing force of the same
amount as that which the sun exerts on the moon, he
simply says,[18] 'the apse of the moon moves about
twice as fast.'

The difficulty of doing what Newton did in this
branch of the subject, and the powers it must have
required, may be judged of from what has already been
stated;—that no one, with his methods, has yet been
able to add anything to his labours: few have undertaken to illustrate what he has written, and no great
number have understood it throughout.   The extreme
complication of the forces, and of the conditions under

---

[15] B. iii. Prop. 23.

[16] B. iii. Prop. 13.

[17] Scholium to Prop. 14. B. iii.

[18] B. i. Prop. 44, second edit.

There is reason to believe, however, that Newton had, in his un-
published calculations, rectified
this discrepancy.

which they act, makes the subject by far the most
thorny walk of mathematics. It is necessary to resolve
the action into many elements, such as can be separated; to invent artifices for dealing with each of
these; and then to recompound the laws thus obtained
into one common conception. The moon's motion
cannot be conceived without comprehending a scheme
more complex than the Ptolemaic epicycles and eccentrics in their worst form; and the component parts of
the system are not, in this instance, mere geometrical
ideas, requiring only a distinct apprehension of relations of space in order to hold them securely; they are
the foundations of mechanical notions, and require to
be grasped so that we can apply to them sound mechanical reasonings. Newton's successors, in the next
generation, abandoned the hope of imitating him in
this intense mental effort; they gave the subject over
to the operation of algebraical reasoning, in which
symbols think for us, without our dwelling constantly
upon their meaning, and obtain for us the consequences
which result from the relations of space and the laws
of force, however complicated be the conditions under
which they are combined. Even Newton's countrymen, though they were long before they applied themselves to the method thus opposed to his, did not produce anything which showed that they had mastered,
or could retrace, the Newtonian investigations.

Thus the Problem of Three Bodies,[19] treated geometrically, belongs exclusively to Newton; and the proofs
of the mutual action of the sun, planets and satellites,
which depend upon such reasoning, could not be discovered by any one but him.

But we have not yet done with his achievements
on this subject; for some of the most remarkable and
beautiful of the reasonings which he connected with
this problem, belong to the next step of his generalization.

5. *Mutual Attraction of all Particles of Matter.*—

---

[19] See the history of the *Problem of Three Bodies, ante,* in Book vi.
Chap. vi. Sect. 7.

That all the parts of the universe are drawn and held together by love, or harmony, or some affection to which, among other names, that of *attraction* may have been given, is an assertion which may very possibly have been made at various times, by speculators writing at random, and taking their chance of meaning and truth. The authors of such casual dogmas have generally nothing accurate or substantial, either in their conception of the general proposition, or in their reference to examples of it; and therefore their doctrines are no concern of ours at present. But among those who were really the first to think of the mutual attraction of matter, we cannot help noticing Francis Bacon; for his notions were so far from being chargeable with the looseness and indistinctness to which we have alluded, that he proposed an experiment[20] which was to decide whether the facts were so or not;— whether the gravity of bodies to the earth arose from an attraction of the parts of matter towards each other, or was a tendency towards the center of the earth. And this experiment is, even to this day, one of the best which can be devised, in order to exhibit the universal gravitation of matter: it consists in the comparison of the rate of going of a clock in a deep mine, and on a high place. Huyghens, in his book *De Causâ Gravitatis,* published in 1690, showed that the earth would have an oblate form, in consequence of the action of the centrifugal force; but his reasoning does not suppose gravity to arise from the mutual attraction of the parts of the earth. The apparent influence of the moon upon the tides had long been remarked; but no one had made any progress in truly explaining the mechanism of this influence; and all the analogies to which reference had been made, on this and similar subjects, as magnetic and other attractions, were rather delusive than illustrative, since they represented the attraction as something peculiar in particular bodies, depending upon the nature of each body.

---

[20] *Nov. Org.* Lib. ii. Aph. 36.

That all such forces, cosmical and terrestrial, were the same single force, and that this was nothing more than the insensible attraction which subsists between one stone and another, was a conception equally bold and grand; and would have been an incomprehensible thought, if the views which we have already explained had not prepared the mind for it. But the preceding steps having disclosed, between all the bodies of the universe, forces of the same kind as those which produce the weight of bodies at the earth, and, therefore, such as exist in every particle of terrestrial matter; it became an obvious question, whether such forces did not also belong to all particles of planetary matter, and whether this was not, in fact, the whole account of the forces of the solar system. But, supposing this conjecture to be thus suggested, how formidable, on first appearance at least, was the undertaking of verifying it! For if this be so, every finite mass of matter exerts forces which are the result of the infinitely numerous forces of its particles, these forces acting in different directions. It does not appear, at first sight, that the law by which the force is related to the distance, will be the same for the particles as it is for the masses; and, in reality, it is not so, except in special cases. And, again, in the instance of any effect produced by the force of a body, how are we to know whether the force resides in the whole mass as a unit, or in the separate particles? We may reason, as Newton does,[21] that the rule which proves gravity to belong universally to the planets, proves it also to belong to their parts; but the mind will not be satisfied with this extension of the rule, except we can find decisive instances, and calculate the effects of both suppositions, under the appropriate conditions. Accordingly, Newton had to solve a new series of problems suggested by this inquiry; and this he did.

These solutions are no less remarkable for the mathematical power which they exhibit, than the other parts of the *Principia*. The propositions in which it

---

[21] *Princip.* B. iii. Prop. 7.

is shown that the law of the inverse square for the
particles gives the same law for spherical masses, have
that kind of beauty which might well have justified
their being published for their mathematical elegance
alone, even if they had not applied to any real case.
Great ingenuity is also employed in other instances,
as in the case of spheroids of small eccentricity. And
when the amount of the mechanical action of masses
of various forms has thus been assigned, the sagacity
shown in tracing the results of such action in the solar
system is truly admirable; not only the general nature
of the effect being pointed out, but its quantity calculated. I speak in particular of the reasonings concerning the Figure of the Earth, the Tides, the Precession of the Equinoxes, the Regression of the Nodes
of a ring such as Saturn's; and of some effects which,
at that time, had not been ascertained even as facts of
observation; for instance, the difference of gravity in
different latitudes, and the Nutation of the earth's
axis. It is true, that in most of these cases, Newton's
process could be considered only as a rude approximation. In one (the Precession) he committed an errour,
and in all, his means of calculation were insufficient.
Indeed these are much more difficult investigations
than the Problem of Three Bodies, in which three
points act on each other by explicit laws. Up to this
day, the resources of modern analysis have been employed upon some of them with very partial success;
and the facts, in all of them, required to be accurately
ascertained and measured, a process which is not completed even now. Nevertheless the form and nature
of the conclusions which Newton did obtain, were such
as to inspire a strong confidence in the competency of
his theory to explain all such phenomena as have been
spoken of. We shall afterwards have to speak of the
labours, undertaken in order to examine the phenomena more exactly, to which the theory gave occasion.

Thus, then, the theory of the universal mutual gravitation of all the particles of matter, according to the
law of the inverse square of the distances, was con-

ceived, its consequences calculated, and its results shown to agree with phenomena. It was found that this theory took up all the facts of astronomy as far as they had hitherto been ascertained; while it pointed out an interminable vista of new facts, too minute or too complex for observation alone to disentangle, but capable of being detected when theory had pointed out their laws, and of being used as criteria or confirmations of the truth of the doctrine. For the same reasoning which explained the evection, variation, and annual equation of the moon, showed that there must be many other inequalities besides these; since these resulted from approximate methods of calculation, in which small quantities were neglected. And it was known that, in fact, the inequalities hitherto detected by astronomers did not give the place of the moon with satisfactory accuracy; so that there was room, among these hitherto untractable irregularities, for the additional results of the theory. To work out this comparison was the employment of the succeeding century; but Newton began it. Thus, at the end of the proposition in which he asserts,[22] that 'all the lunar motions and their irregularities follow from the principles here stated,' he makes the observation which we have just made; and gives, as examples, the different motions of the apogee and nodes, the difference of the change of the eccentricity, and the difference of the moon's variation, according to the different distances of the sun. ' But this inequality,' he says, ' in astronomical calculations, is usually referred to the prosthaphæresis of the moon, and confounded with it.'

Reflections on the Discovery.—Such, then, is the great Newtonian Induction of Universal Gravitation, and such its history. It is indisputably and incomparably the greatest scientific discovery ever made, whether we look at the advance which it involved, the extent of the truth disclosed, or the fundamental and satisfactory nature of this truth. As to the first point,

---

[22] B. iii. Prop. 22.

we may observe that any one of the five steps into which we have separated the doctrine, would, of itself, have been considered as an important advance;—would have conferred distinction on the persons who made it, and the time to which it belonged. All the five steps made at once, formed not a leap, but a flight,—not an improvement merely, but a metamorphosis,—not an epoch, but a termination. Astronomy passed at once from its boyhood to mature manhood. Again, with regard to the extent of the truth, we obtain as wide a generalization as our physical knowledge admits, when we learn that every particle of matter, in all times, places, and circumstances, attracts every other particle in the universe by one common law of action. And by saying that the truth was of a fundamental and satisfactory nature, I mean that it assigned, not a rule merely, but a cause, for the heavenly motions; and that kind of cause which most eminently and peculiarly we distinctly and thoroughly conceive, namely, mechanical force. Kepler's laws were merely *formal* rules, governing the celestial motions according to the relations of space, time, and number; Newton's was a *causal* law, referring these motions to mechanical reasons. It is no doubt conceivable that future discoveries may both extend and further explain Newton's doctrines;—may make gravitation a case of some wider law, and may disclose something of the mode in which it operates; questions with which Newton himself struggled. But, in the mean time, few persons will dispute, that both in generality and profundity, both in width and depth, Newton's theory is altogether without a rival or neighbour.[23]

---

[23] The value and nature of this step have long been generally acknowledged wherever science is cultivated. Yet it would appear that there is, in one part of Europe, a school of philosophers who contest the merit of this part of Newton's discoveries. 'Kepler,' says a celebrated German metaphysician,* 'discovered the laws of free motion; a discovery of immortal glory. It has since been the fashion to say that Newton first found out the proof of these

* Hegel, *Encyclopædia*, § 270.

The requisite conditions of such a discovery in the mind of its author were, in this as in other cases, the idea, and its comparison with facts;—the conception of the law, and the moulding this conception in such a form as to correspond with known realities. The idea of mechanical force as the cause of the celestial motions, had, as we have seen, been for some time growing up in men's minds;—had gone on becoming more distinct and more general; and had, in some persons, approached the form in which it was entertained by Newton. Still, in the mere conception of universal gravitation, Newton must have gone far beyond his predecessors and contemporaries, both in generality and distinctness; and in the inventiveness and sagacity with which he traced the consequences of this conception, he was, as we have shown, without a rival, and almost without a second. As to the facts which he had to include in his law, they had been accumulating from the very birth of astronomy; but those which he had more peculiarly to take hold of, were the facts of the planetary motions as given by Kepler, and those of the moon's motions as given by Tycho Brahe and Jeremy Horrox.

We find here occasion to make a remark which is important in its bearing on the nature of progressive science. What Newton thus used and referred to as *facts*, were the *laws* which his predecessors had esta-

---

rules. It has seldom happened that the glory of the first discoverer has been more unjustly transferred to another person.' It may appear strange that any one in the present day should hold such language; but if we examine the reasons which this author gives, they will be found, I think, to amount to this; that his mind is in the condition in which Kepler's was; and that the whole range of mechanical ideas and modes of conception which made the transition from Kepler to Newton possible, are extraneous to the domain of his philosophy. Even this author, however, if I understand him rightly, recognizes Newton as the author of the doctrine of Perturbations.

I have given a further account of these views, in a Memoir *On Hegel's Criticism of Newton's Principia*. Cambridge Transactions, 1849.

blished. What Kepler and Horrox had put forth as 'theories,' were now established truths, fit to be used in the construction of other theories. It is in this manner that one theory is built upon another;—that we rise from particulars to generals, and from one generalization to another;—that we have, in short, successive steps of induction. As Newton's laws assumed Kepler's, Kepler's laws assumed as facts the results of the planetary theory of Ptolemy; and thus the theories of each generation in the scientific world are (when thoroughly verified and established) the facts of the next generation. Newton's theory is the circle of generalization which includes all the others; —the highest point of the inductive ascent;—the catastrophe of the philosophic drama to which Plato had prologized;—the point to which men's minds had been journeying for two thousand years.

Character of Newton.—It is not easy to anatomize the constitution and the operations of the mind which makes such an advance in knowledge. Yet we may observe that there must exist in it, in an eminent degree, the elements which compose the mathematical talent. It must possess distinctness of intuition, tenacity and facility in tracing logical connexion, fertility of invention, and a strong tendency to generalization. It is easy to discover indications of these characteristics in Newton. The distinctness of his intuitions of space, and we may add of force also, was seen in the amusements of his youth; in his constructing clocks and mills, carts and dials, as well as the facility with which he mastered geometry. This fondness for handicraft employments, and for making models and machines, appears to be a common prelude of excellence in physical science;[24] probably on this very account, that it arises from the distinctness of intuitive power with which the child conceives the shapes and the working of such material combinations. Newton's inventive power appears in the number and

---

[24] As in Galileo, Hooke, Huyghens, and others.

variety of the mathematical artifices and combinations which he devised, and of which his books are full. If we conceive the operation of the inventive faculty in the only way in which it appears possible to conceive it;—that while some hidden source supplies a rapid stream of possible suggestions, the mind is on the watch to seize and detain any one of these which will suit the case in hand, allowing the rest to pass by and be forgotten;—we shall see what extraordinary fertility of mind is implied by so many successful efforts; what an innumerable host of thoughts must have been produced, to supply so many that deserved 'to be selected. And since the selection is performed by tracing the consequences of each suggestion, so as to compare them with the requisite conditions, we see also what rapidity and certainty in drawing conclusions the mind must possess as a talent, and what watchfulness and patience as a habit.

The hidden fountain of our unbidden thoughts is for us a mystery; and we have, in our consciousness, no standard by which we can measure our own talents; but our acts and habits are something of which we are conscious; and we can understand, therefore, how it was that Newton could not admit that there was any difference between himself and other men, except in his possession of such habits as we have mentioned, perseverance and vigilance. When he was asked how he made his discoveries, he answered, ' by always thinking about them;' and at another time he declared that if he had done anything, it was due to nothing but industry and patient thought: 'I keep the subject of my inquiry constantly before me, and wait till the first dawning opens gradually, by little and little, into a full and clear light.' No better account can be given of the nature of the mental *effort* which gives to the philosopher the full benefit of his powers; but the natural *powers* of men's minds are not on that account the less different. There are many who might wait through ages of darkness without being visited by any dawn.

The habit to which Newton thus, in some sense,

owed his discoveries, this constant attention to the rising thought, and development of its results in every direction, necessarily engaged and absorbed his spirit, and made him inattentive and almost insensible to external impressions and common impulses. The stories which are told of his extreme absence of mind, probably refer to the two years during which he was composing his *Principia,* and thus following out a train of reasoning the most fertile, the most complex, and the most important, which any philosopher had ever had to deal with. The magnificent and striking questions which, during this period, he must have had daily rising before him; the perpetual succession of difficult problems of which the solution was necessary to his great object; may well have entirely occupied and possessed him. ' He existed only to calculate and to think.'[25] Often, lost in meditation, he knew not what he did, and his mind appeared to have quite forgotten its connexion with the body. His servant reported that, in rising in a morning, he frequently sat a large portion of the day, half-dressed, on the side of his bed; and that his meals waited on the table for hours before he came to take them. Even with his transcendent powers, to do what he did, was almost irreconcileable with the common conditions of human life; and required the utmost devotion of thought, energy of effort, and steadiness of will,—the strongest character, as well as the highest endowments, which belong to man.

Newton has been so universally considered as the greatest example of a natural philosopher, that his moral qualities, as well as his intellect, have been referred to as models of the philosophical character; and those who love to think that great talents are naturally associated with virtue, have always dwelt with pleasure upon the views given of Newton by his contemporaries; for they have uniformly represented him as candid and humble, mild and good. We may take as an example of the impressions prevalent about

---

[25] Biot.

him in his own time, the expressions of Thomson, in the Poem on his Death.[26]

> Say ye who best can tell, ye happy few,
> Who saw him in the softest lights of life,
> All unwithheld, indulging to his friends
> The vast unborrowed treasures of his mind,
> Oh, speak the wondrous man! how mild, how calm,
> How greatly humble, how divinely good,
> How firm established on eternal truth!
> Fervent in doing well, with every nerve
> Still pressing on, forgetful of the past,
> And panting for perfection; far above
> Those little cares and visionary joys
> That so perplex the fond impassioned heart
> Of ever-cheated, ever-trusting man.

[2nd Ed.] [In the first edition of the *Principia*, published in 1687, Newton showed that the nature of all the then known inequalities of the moon, and in some cases their quantities, might be deduced from the principles which he laid down: but the determination of the amount and law of most of the inequalities was deferred to a more favourable opportunity, when he might be furnished with better astronomical observations. Such observations as he needed for this purpose had been made by Flamsteed, and for these he applied, representing how much value their use would add to the observations. 'If,' he says, in 1694, 'you publish them without such a theory to recommend them, they will only be thrown into the heap of the observations of former astronomers, till somebody shall arise that by perfecting the theory of the moon shall discover your observations to be exacter than the rest;

---

[26] In the same strain we find the general voice of the time. For instance, one of Loggan's ' Views of Cambridge' is dedicated 'Isaaco Newtono.. Mathematico, Physico, Chymico consummatissimo; nec minus suavitate morum et candore animi ... spectabili.'

In opposition to the general current of such testimony, we have the complaints of Flamsteed, who ascribes to Newton angry language and harsh conduct in the matter of the publication of the Greenwich Observations, and of Whiston. Yet even Flamsteed speaks well of his general disposition. Whiston was himself so weak and prejudiced that his testimony is worth very little.

but when that shall be, God knows : I fear, not in your lifetime, if I should die before it is done. For I find this theory so very intricate, and the theory of gravity so necessary to it, that I am satisfied it will never be perfected but by somebody who understands the theory of gravity as well, or better than I do.' He obtained from Flamsteed the lunar observations for which he applied, and by using these he framed the Theory of the Moon which is given as his in David Gregory's *Astronomiæ Elementa.*[27] He also obtained from Flamsteed the diameters of the planets as observed at various times, and the greatest elongations of Jupiter's Satellites, both of which, Flamsteed says, he made use of in his *Principia.*

Newton, in his letters to Flamsteed in 1694 and 5, acknowledges this service.[28]]

---

[27] In the Preface to a *Treatise on Dynamics*, Part i., published in 1836, I have endeavoured to show that Newton's modes of determining several of the lunar inequalities admitted of an accuracy not very inferior to the modern analytical methods.

[28] The quarrel on the subject of the publication of Flamsteed's Observations took place at a later period. Flamsteed wished to have his Observations printed complete and entire. Halley, who, under the authority of Newton and others, had the management of the printing, made many alterations and omissions, which Flamsteed considered as deforming and spoiling the work. The advantages of publishing a *complete* series of observations, now generally understood, were not then known to astronomers in general, though well known to Flamsteed, and earnestly insisted upon in his remonstrances. The result was that Flamsteed published his Observations at his own expense, and finally obtained permission to destroy the copies printed by Halley, which he did. In 1726, after Flamsteed's death, his widow applied to the Vice-Chancellor of Oxford, requesting that the volume printed by Halley might be removed out of the Bodleian Library, where it exists, as being ' nothing more than an erroneous abridgement of Mr. Flamsteed's works,' and unfit to see the light.

INTRODUCTION.

Of Thermotics and Atmology.

I EMPLOY the term *Thermotics*, to include all the doctrines respecting Heat, which have hitherto been established on proper scientific grounds. Our survey of the history of this branch of science must be more rapid and less detailed than it has been in those subjects of which we have hitherto treated: for our knowledge is, in this case, more vague and uncertain than in the others, and has made less progress towards a general and certain theory. Still, the narrative is too important and too instructive to be passed over.

The distinction of Formal Thermotics and Physical Thermotics,—of the discovery of the mere Laws of Phenomena, and the discovery of their Causes,—is applicable here, as in other departments of our knowledge. But we cannot exhibit, in any prominent manner, the latter division of the science now before us; since no general theory of heat has yet been propounded, which affords the means of calculating the circumstances of the phenomena of conduction, radiation, expansion, and change of solid, liquid, and gaseous form. Still, on each of these subjects there have been proposed, and extensively assented to, certain general views, each of which explains its appropriate class of phenomena; and, in some cases, these principles have been clothed in precise and mathematical conditions, and thus made bases of calculation.

These principles, thus possessing a generality of a limited kind, connecting several observed laws of phenomena, but yet not connecting all the observed classes of facts which relate to heat, will require our separate attention. They may be described as the Doctrine of Conduction, the Doctrine of Radiation, the Doctrine of Specific Heat, and the Doctrine of Latent Heat; and these, and similar doctrines respecting heat, make up the science which we may call *Thermotics proper.*

But besides these collections of principles which regard heat by itself, the relations of heat and moisture give rise to another extensive and important collection of laws and principles, which I shall treat of in connexion with Thermotics, and shall term *Atmology,* borrowing the term from the Greek word (ἄτμος,) which signifies *vapour.* The *Atmosphere* was so named by the Greeks, as being a sphere of vapour; and, undoubtedly, the most general and important of the phenomena which take place in the air, by which the earth is surrounded, are those in which water, of one *consistence* or other (ice, water, or steam), is concerned. The knowledge which relates to what takes place in the atmosphere has been called *Meteorology,* in its collective form: but such knowledge is, in fact, composed of parts of many different sciences. And it is useful for our purpose to consider separately those portions of Meteorology which have reference to the laws of aqueous vapour, and these we may include under the term Atmology.

The instruments which have been invented for the purpose of measuring the moisture of the air, that is, the quantity of vapour which exists in it, have been termed *Hygrometers;* and the doctrines on which these instruments depend, and to which they lead, have been called *Hygrometry;* but this term has not been used in quite so extensive a sense as that which we intend to affix to *Atmology.*

In treating of Thermotics, we shall first describe the earlier progress of men's views concerning Conduction, Radiation, and the like, and shall then speak of the more recent corrections and extensions, by which they have been brought nearer to theoretical generality.

CHAPTER IV.

PHYSICAL THEORIES OF HEAT.

WHEN we look at the condition of that branch of knowledge which, according to the phraseology already employed, we must call *Physical Thermotics*, in opposition to Formal Thermotics, which gives us detached laws of phenomena, we find the prospect very different from that which was presented to us by physical astronomy, optics, and acoustics. In these sciences, the maintainers of a distinct and comprehensive theory have professed at least to show that it explains and includes the principal laws of phenomena of very various kinds: in Thermotics, we have only attempts to explain a part of the facts. We have here no example of an hypothesis which, assumed in order to explain one class of phenomena, has been found also to account exactly for another; as when central forces led to the precession of the equinoxes; or when the explanation of polarization explained also double refraction; or when the pressure of the atmosphere, as measured by the barometer, gave the true velocity of sound. Such coincidences, or *consiliences*, as I have elsewhere called them, are the test of truth; and thermotical theories cannot yet exhibit credentials of this kind.

On looking back at our view of this science, it will be seen that it may be distinguished into two parts; the Doctrines of Conduction and Radiation, which we call Thermotics proper; and the Doctrines respecting the relation of Heat, Airs, and Moisture, which we have termed Atmology. These two subjects differ in their bearing on our hypothetical views.

Thermotical Theories.—The phenomena of radiant heat, like those of radiant light, obviously admit of general explanation in two different ways;—by the emission of material particles, or by the propagation of undulations. Both these opinions have found supporters. Probably most persons, in adopting Prevost's theory of exchanges, conceive the radiation of heat to be the radiation of matter. The undulation hypothesis, on the other hand, appears to be suggested by the production of heat by friction, and was accordingly maintained by Rumford and others. Leslie[1] appears, in a great part of his *Inquiry*, to be a supporter of some undulatory doctrine, but it is extremely difficult to make out what his undulating medium is; or rather, his opinions wavered during his progress. In page 31, he asks, 'What is this calorific and frigorific fluid?' and after keeping the reader in suspense for a moment, he replies,

'Quod petis hic est.

It is merely the ambient AIR.' But at page 150, he again asks the question, and, at page 188, he answers, 'It is the same subtile matter that, according to its different modes of existence, constitutes either heat or light.' A person thus vacillating between two opinions, one of which is palpably false, and the other laden with exceeding difficulties which he does not even attempt to remove, had little right to protest against[2] 'the sportive freaks of some intangible *aura*;' to rank all other hypotheses than his own with the 'occult qualities of the schools;' and to class the 'prejudices' of his opponents with the tenets of those who maintained the *fuga vacui* in opposition to Torricelli. It is worth while noticing this kind of rhetoric, in order to observe, that it may be used just as easily on the wrong side as on the right.

Till recently, the theory of material heat, and of its propagation by emission, was probably the one most in favour with those who had studied mathematical thermotics. As we have said, the laws of conduction, in their ultimate analytical form, were almost identical with the laws of motion of fluids. Fourier's principle also, that the radiation of heat takes place from points below the surface, and is intercepted by the super-

---

[1] *An Experimental Inquiry into the Nature and Propagation of Heat.* 1804.   [2] Ib. p. 47.

ficial particles, appears to favour the notion of material emission.

Accordingly, some of the most eminent modern French mathematicians have accepted and extended the hypothesis of a material caloric. In addition to Fourier's doctrine of molecular extra-radiation, Laplace and Poisson have maintained the hypothesis of *molecular intra-radiation*, as the mode in which conduction takes place: that is, they say that the particles of bodies are to be considered as *discrete*, or as points separated from each other, and acting on each other at a distance; and the conduction of heat from one part to another, is performed by radiation between all neighbouring particles. They hold that, without this hypothesis, the differential equations expressing the conditions of conduction cannot be made homogeneous: but this assertion rests, I conceive, on an errour, as Fourier has shown, by dispensing with the hypothesis. The necessity of the hypothesis of discrete molecular action in bodies, is maintained in all cases by M. Poisson; and he has asserted Laplace's theory of capillary attraction to be defective on this ground, as Laplace asserted Fourier's reasoning respecting heat to be so. In reality, however, this hypothesis of discrete molecules cannot be maintained as a physical truth; for the law of molecular action, which is assumed in the reasoning, after answering its purpose in the progress of calculation, vanishes in the result; the conclusion is the same, whatever law of the intervals of the molecules be assumed. The definite integral, which expresses the whole action, no more proves that this action is actually made of the differential parts by means of which it was found, than the processes of finding the weight of a body by integration, prove it to be made up of differential weights. And therefore, even if we were to adopt the emission theory of heat, we are by no means bound to take along with it the hypothesis of discrete molecules.

But the recent discovery of the refraction, polarization, and depolarization of heat, has quite altered the theoretical aspect of the subject, and, almost, at a single blow, ruined the emission theory. Since heat is reflected and refracted like light, analogy would lead us to conclude that the mechanism of the processes is the same in the two cases. And when we add to these properties the property of polarization, it is scarcely possible to believe otherwise than that heat consists in transverse vibrations; for no wise philosopher would attempt an explanation by ascribing poles to the emitted particles, after the experience which Optics affords, of the utter failure of such machinery.

But here the question occurs, If heat consists in vibrations, whence arises the extraordinary identity of the laws of its propagation with the laws of the flow of matter? How is it that, in conducted heat, this vibration creeps slowly from one part of the body to another, the part first heated remaining hottest; instead of leaving its first place and travelling rapidly to another, as the vibrations of sound and light do? The answer to these questions has been put in a very distinct and plausible form by that distinguished philosopher, M. Ampère, who published a *Note on Heat and Light considered as the results of Vibratory Motion*,[3] in 1834 and 1835; and though this answer is an hypothesis, it at least shows that there is no fatal force in the difficulty.

M. Ampère's hypothesis is this; that bodies consist of solid molecules, which may be considered as arranged at intervals in a very rare ether; and that the vibrations of the molecules, causing vibrations of the ether and caused by them, constitute heat. On these suppositions, we should have the phenomena of conduction explained; for if the molecules at one end of a bar be hot, and therefore in a state of vibration, while the others are at rest, the vibrating molecules propagate vibrations in the ether, but these vibrations do not produce heat, except in proportion as they put the quiescent molecules of the bar in vibration; and the ether being very rare compared with the molecules, it is only by the

---

[3] *Bibliothèque Universelle de Genève*, vol. xlix. p. 225.   *Ann. Chim.* tom. lvii. p. 434.

repeated impulses of many successive vibrations that the nearest quiescent molecules are made to vibrate; after which they combine in communicating the vibration to the more remote molecules. 'We then find necessarily,' M. Ampère adds, 'the same equations as those found by Fourier for the distribution of heat, setting out from the same hypothesis, that the temperature or heat transmitted is proportional to the difference of the temperatures.'

Since the undulatory hypothesis of heat can thus answer all obvious objections, we may consider it as upon its trial, to be confirmed or modified by future discoveries; and especially, by an enlarged knowledge of the laws of the polarization of heat.

[2nd Ed.] [Since the first edition was written, the analogies between light and heat have been further extended, as I have already stated. It has been discovered by MM. Biot and Melloni that quartz impresses a circular polarization upon heat; and by Prof. Forbes that mica, of a certain thickness, produces phenomena such as would be produced by the impression of circular polarization on the supposed transversal vibrations of radiant heat; and further, a rhomb of rock-salt, of the shape of the glass rhomb which verified Fresnel's extraordinary anticipation of the circular polarization of light, verified the expectation, founded upon other analogies, of the polarization of heat. By passing polarized heat through various thicknesses of mica, Prof. Forbes has attempted to calculate the length of an undulation for heat.

These analogies cannot fail to produce a strong disposition to believe that light and heat, essences so closely connected that they can hardly be separated, and thus shown to have so many curious properties in common, are propagated by the same machinery; and thus we are led to an Undulatory Theory of Heat.

Yet such a Theory has not yet by any means received full confirmation. It depends upon the analogy and the connexion of the Theory of Light, and would have little weight if those were removed. For, the separa-

F F

tion of the rays in double refraction, and the phenomena of periodical intensity, the two classes of facts out of which the Undulatory Theory of Optics principally grew, have neither of them been detected in thermotical experiments. Prof. Forbes has assumed alternations of heat for increasing thicknesses of mica, but in his experiments we find only one *maximum*. The occurrence of alternate maxima and minima under the like circumstances would exhibit visible waves of heat, as the fringes of shadows do of light, and would thus add much to the evidence of the theory.

Even if I conceived the Undulatory Theory of Heat to be now established, I should not venture, as yet, to describe its establishment as an event in the history of the Inductive Sciences. It is only at an interval of time after such events have taken place that their history and character can be fully understood, so as to suggest lessons in the Philosophy of Science.]

Atmological Theories.—Hypotheses of the relations of heat and air almost necessarily involve a reference to the forces by which the composition of bodies is produced, and thus cannot properly be treated of till we have surveyed the condition of chemical knowledge. But we may say a few words on one such hypothesis; I mean the hypothesis on the subject of the atmological laws of heat, proposed by Laplace, in the twelfth Book of the *Mécanique Céleste*, and published in 1823. It will be recollected that the main laws of phenomena for which we have to account, by means of such an hypothesis, are the following:—

(1.) The Law of Boyle and Mariotte, that the elasticity of an air varies as its density. See Chap. iii. Sect. 1 of this Book.

(2.) The Law of Gay-Lussac and Dalton, that all airs expand equally by heat. See Chap. ii. Sect. 1.

(3.) The production of heat by sudden compression. See Chap. ii. Sect. 2.

(4.) Dalton's principle of the mechanical mixture of airs. See Chap. iii. Sect. 3.

(5.) The Law of expansion of solids and fluids by heat. See Chap. ii. Sect. 1.

(6.) Changes of consistence by heat, and the doctrine of latent heat. See Chap. ii. Sect. 3.

(7.) The Law of the expansive force of steam. See Chap. iii. Sect. 4.

Besides these, there are laws of which it is doubtful whether they are or are not included in the preceding, as the low temperature of the air in the higher parts of the atmosphere. (See Chap. iii. Sect. 5.)

Laplace's hypothesis[4] is this:—that bodies consist of particles, each of which gathers round it, by its attraction, a quantity of caloric: that the particles of the bodies attract each other, besides attracting the caloric, and that the particles of the caloric repel each other.

In gases, the particles of the bodies are so far removed, that their mutual attraction is insensible, and the matter tends to expand by the mutual repulsion of the caloric. He conceives this caloric to be constantly radiating among the particles; the density of this internal radiation is the *temperature*, and he proves that, on this supposition, the elasticity of the air will be as the density, and as this temperature. Hence follow the three first rules above stated. The same suppositions lead to Dalton's principle of mixtures (4), though without involving his mode of conception; for Laplace says that whatever the mutual action of two gases be, the whole pressure will be equal to the sum of the separate pressures.[5] Expansion (5), and the changes of consistence (6), are explained by supposing[6] that in solids, the mutual attraction of the particles of the body is the greatest force; in liquids, the attraction of the particles for the caloric; in airs, the repulsion of the caloric. But the doctrine of latent heat again modifies[7] the hypothesis, and makes it necessary to include latent heat in the calculation; yet there is not, as we might hope there would be if the theory were the true one, any confirmation of the

---

[4] *Méc. Cél.* t. v. p. 89.   [5] Ib. p. 110.   [6] Ib. p. 92.
[7] *Méc. Cél.* t. v. p. 93.

hypothesis resulting from the new class of laws thus referred to. Nor does it appear that the hypothesis accounts for the relation between the elasticity and the temperature of steam.

It will be observed that Laplace's hypothesis goes entirely upon the materiality of heat, and is inconsistent with any vibratory theory; for, as Ampère remarks, 'It is clear that if we admit heat to consist in vibrations, it is a contradiction to attribute to heat (or caloric) a repulsive force of the particles which would be a cause of vibration.'

An unfavourable judgment of Laplace's Theory of Gases is suggested by looking for that which, in speaking of Optics, was mentioned as the great characteristic of a true theory; namely, that the hypotheses, which were assumed in order to account for one class of facts, are found to explain another class of a different nature:—the consilience of inductions. Thus, in thermotics, the law of an intensity of radiation proportional to the sine of the angle of the ray with the surface, which is founded on direct experiments of radiation, is found to be necessary in order to explain the tendency of neighbouring bodies to equality of temperature; and this leads to the higher generalization, that heat is radiant from points below the surface. But in the doctrine of the relation of heat to gases, as delivered by Laplace, there is none of this unexpected confirmation; and though he explains some of the leading laws, his assumptions bear a large proportion to the laws explained. Thus, from the assumption that the repulsion of gases arises from the mutual repulsion of the particles of caloric, he finds that the pressure in any gas is as the square of the density and of the quantity of caloric;[8] and from the assumption that the temperature is the internal radiation, he finds that this temperature is as the density and the square of the caloric.[9] Hence he obtains the law of Boyle and Mariotte, and that of Dalton and Gay-Lussac. But this view of the subject

---

[8] $P = 2 \pi H K \rho^2 c^2$ (1) p. 107.   [9] $q' \Pi (a) = \rho c^2$ (2) p. 108.

requires other assumptions when we come to latent heat; and accordingly, he introduces, to express the latent heat, a new quantity.[10]   Yet this quantity produces no effect on his calculations, nor does he apply his reasoning to any problem in which latent heat is concerned.

Without, then, deciding upon this theory, we may venture to say that it is wanting in all the prominent and striking characteristics which we have found in those great theories which we look upon as clearly and indisputably established.

Conclusion.—We may observe, moreover, that heat has other bearings and effects, which, as soon as they have been analysed into numerical laws of phenomena, must be attended to in the formation of thermotical theories.   Chemistry will probably supply many such; those which occur to us, we must examine hereafter. But we may mention as examples of such, MM. De la Rive and Marcet's law, that the specific heat of all gases is the same;[11] and MM. Dulong and Petit's law, that single atoms of all simple bodies have the same capacity for heat.[12]   Though we have not yet said anything of the relation of different gases, or explained the meaning of *atoms* in the chemical sense, it will easily be conceived that these are very general and important propositions.

Thus the science of Thermotics, imperfect as it is, forms a highly-instructive part of our survey; and is one of the cardinal points on which the doors of those chambers of physical knowledge must turn which hitherto have remained closed.   For, on the one hand, this science is related by strong analogies and dependencies to the most complete portions of our knowledge, our mechanical doctrines and optical theories; and on the other, it is connected with properties and laws of a nature altogether different,—those of chemistry; properties and laws depending upon a new

---

[10] The quantity *i*, p. 113.

[11] *Ann. Chim.* xxxv. (1827.)   [12] Ib. x. 397.

system of notions and relations, among which clear and substantial general principles are far more difficult to lay hold of, and with which the future progress of human knowledge appears to be far more concerned. To these notions and relations we must now proceed; but we shall find an intermediate stage, in certain subjects which I shall call the *Mechanico-chemical* Sciences; viz., those which have to do with Magnetism, Electricity, and Galvanism.

CHAPTER IV.

Theories of Heat.

The Dynamical Theory of Heat.

THAT the transmission of *radiant* Heat takes place by means of the vibrations of a medium, as the transmission of Sound certainly does, and the transmission of Light most probably, is a theory which, as I have endeavoured to explain, has strong arguments and analogies in its favour. But that Heat itself, in its essence and quantity, is Motion, is a hypothesis of quite another kind. This hypothesis has been recently asserted and maintained with great ability. The doctrine thus asserted is, that Motion may be converted into Heat, and Heat into Motion; that Heat and Motion may produce each other, as we see in the rarefaction and condensation of air, in steam-engines, and the like: and that in all such cases the Motion produced and the Heat expended exactly measure each other. The foundation of this theory is conceived to have been laid by Mr. Joule of Manchester, in 1844; and it has since been prosecuted by him and by Professor Thomson of Glasgow, by experimental investigations of various kinds. It is difficult to make these experiments so as to be quite satisfactory; for it is difficult to measure *all* the heat gained or lost in any of the changes here contemplated. That friction, agitation of fluids, condensation of gases, conversion of gases into fluids and liquids into solids, produce heat, is undoubted: and that the quantity of such heat may be measured by the mechanical force which produces it, or which it produces, is a generalization which will very likely be found a fertile source of new propositions, and probably of important consequences.

As an example of the conclusions which Professor Thomson draws from this doctrine of the mutual conversion of motion and heat, I may mention his speculations concerning the cause which produces and sustains the heat of the sun.[1] He conceives that the support of the solar heat must be meteoric matter which is perpetually falling towards the globe of the sun, and has its motion converted into heat. He inclines to think that the meteors containing the stores of energy for future Sun-light must be principally within the earth's orbit; and that we actually see them there as the 'Zodiacal Light,' an illuminated shower, or rather tornado, of stones. The inner parts of this tornado are always getting caught in the Sun's atmosphere, and drawn to his mass by gravitation.

---

[1] Of the Mechanical Energies of the Solar System. *Edinb. Trans.* vol. xxi. part i. (1854), p. 67.

END OF VOL. II.

INTRODUCTION.

Sect. 1.—*Of the Classificatory Sciences.*

THE horizon of the sciences spreads wider and wider before us, as we advance in our task of taking a survey of the vast domain. We have seen that the existence of Chemistry as a science which declares the ingredients and essential constitution of all kinds of bodies, implies the existence of another corresponding science, which shall divide bodies into kinds, and point out steadily and precisely what bodies they are which we have analysed. But a science thus dividing and defining bodies, is but one member of an order of sciences, different from those which we have hitherto described; namely, of the *classificatory sciences.* Such sciences there must be, not only having reference to the bodies with which chemistry deals, but also to all things respecting which we aspire to obtain any general knowledge, as, for instance, plants and animals. Indeed it will be found, that it is with regard to these latter objects, to organized beings, that the process of scientific classification has been most successfully exercised; while with regard to inorganic substances, the formation of a satisfactory system of arrangement has been found extremely difficult; nor has the necessity of such a system been recognized by chemists so distinctly and constantly as it ought to be. The best exemplifications of these branches of knowledge, of which we now have to speak, will, therefore, be found in the organic world, in Botany and Zoology; but we will, in the first place, take a brief view of the science which classifies inorganic bodies, and of which Mineralogy is hitherto the very imperfect representative.

The principles and rules of the Classificatory Sciences, as well as of those of the other orders of sciences, must be fully explained when we come to treat of the Philosophy of the Sciences; and cannot be introduced here, where we have to do with history only. But I may observe very briefly, that with the process of *classing*, is joined the process of *naming;*—that names imply classification;—and that even the rudest and earliest application of language presupposes a distribution of objects according to their kinds;—but that such a spontaneous and unsystematic distribution cannot, in the cases we now have to consider, answer the purposes of exact and general knowledge. Our classification of objects must be made consistent and systematic, in order to be scientific; we must discover marks and characters, properties and conditions, which are constant in their occurrence and relations; we must form our classes, we must impose our names, according to such marks. We can thus, and thus alone, arrive at that precise, certain, and systematic knowledge, which we seek; that is, at science. The object, then, of the classificatory sciences is to obtain FIXED CHARACTERS of the kinds of things; and the criterion of the fitness of names is, that THEY MAKE GENERAL PROPOSITIONS POSSIBLE.

I proceed to review the progress of certain sciences on these principles, and first, though briefly, the science of Mineralogy.

Sect. 2.—Of Mineralogy as the Analytico-classificatory Science.

MINERALOGY, as it has hitherto been cultivated, is, as I have already said, an imperfect representative of the department of human knowledge to which it belongs. The attempts at the science have generally been made by collecting various kinds of information respecting mineral bodies; but the science which we require is a complete and consistent classified system of all inorganic bodies. For chemistry proceeds upon the principle that the constitution of a body invariably determines its properties; and, consequently, its kind: but we cannot apply this principle, except we can speak with precision of the *kind* of a body, as well as of its composition. We cannot attach any sense to the assertion, that 'soda or baryta has a metal for its base,' except we know what *a metal* is, or at least what properties it implies. It may not be, indeed it is not, possible, to define the kinds of bodies by words only; but the classification must proceed by some constant and generally applicable process; and the knowledge which has reference to the classification will be precise as far as this process is precise, and vague as far as this is vague.

There must be, then, as a necessary supplement to Chemistry, a Science of those properties of bodies by which we divide them into *kinds*. Mineralogy is the branch of knowledge which has discharged the office of such a science, so far as it has been discharged; and, indeed, Mineralogy has been gradually approaching to a clear consciousness of her real place, and of her whole task; I shall give the history of some of the advances which have thus been made. They are, principally, the establishment and use of External Characters, especially of *Crystalline Form*, as a fixed character of definite substances; and the attempts to bring into view the connexion of Chemical Constitution and External Properties, made in the shape of mineralogical *Systems;* both those in which *chemical methods of arrangement* are adopted, and those which profess to classify by the *natural-history method.*

CRYSTALLOGRAPHY.

CHAPTER I.

Prelude to the Epoch of De Lisle and Haüy.

OF all the physical properties of bodies, there is none so fixed, and in every way so remarkable, as this;—that the same chemical compound always assumes, with the utmost precision, the same geometrical form. This identity, however, is not immediately obvious; it is often obscured by various mixtures and imperfections in the substance; and even when it is complete, it is not immediately recognized by a common eye, since it consists, not in the equality of the sides or faces of the figures, but in the equality of their angles. Hence it is not surprising that the constancy of form was not detected by the early observers. Pliny says,[1] 'Why crystal is generated in a hexagonal form, it is difficult to assign a reason; and the more so, since, while its faces are smoother than any art could make them, the pyramidal points are *not all of the same kind.*' The quartz crystals of the Alps, to which he refers, are, in some specimens, very regular, while in others, one side of the pyramid becomes much the largest; yet the angles remain constantly the same. But when the whole shape varied so much, the angles also seemed to vary. Thus Conrad Gessner. a very learned naturalist, who, in 1564, published at Zurich his work, *De rerum Fossilium, Lapidum et Gemmarum maxime, Figuris,* says,[2] 'One crystal differs from another in its angles, and consequently in its figure.' And Cæsalpinus, who, as we shall find, did so much in establishing fixed characters in botany, was led by some of his general views to

---

[1] *Nat. Hist.* xxvii. 2.   [2] p. 25.

disbelieve the fixity of the form of crystals. In his work *De Metallicis*, published at Nuremberg in 1602, he says,[3] 'To ascribe to inanimate bodies a definite form, does not appear consentaneous to reason; for it is the office of organization to produce a definite form;' an opinion very natural in one who had been immersed in the study of the general analogies of the forms of plants. But though this is excusable in Cæsalpinus, the rejection of this definiteness of form a hundred and eighty years later, when its existence had been proved, and its laws developed by numerous observers, cannot be ascribed to anything but strong prejudice; yet this was the course taken by no less a person than Buffon. 'The form of crystallization,' says he,[4] 'is *not a constant character*, but is more equivocal and more variable than any other of the characters by which minerals are to be distinguished.' And accordingly, he makes no use of this most important feature in his history of minerals. This strange perverseness may perhaps be ascribed to the dislike which Buffon is said to have entertained for Linnæus, who had made crystalline form a leading character of minerals.

It is not necessary to mark all the minute steps by which mineralogists were gradually led to see clearly the nature and laws of the fixity of crystalline forms. These forms were at first noticed in that substance which is peculiarly called rock-crystal or quartz; and afterwards in various stones and gems, in salts obtained from various solutions, and in snow. But those who observed the remarkable regular figures which these substances assume, were at first impelled onwards in their speculations by the natural tendency of the human mind to generalize and guess, rather than to examine and measure. They attempted to snatch at once the general laws of geometrical regularity of these occurrences, or to connect them with some doctrine concerning formative causes. Thus Kepler,[5] in his *Harmonics of the World*, asserts a '*formatrix facultas*, which has its seat in the entrails of the earth, and,

---

[3] p. 97.   [4] *Hist. des Min.* p. 343.   [5] Linz. 1619, p. 161.

after the manner of a pregnant woman, expresses the five regular geometrical solids in the forms of gems.' But philosophers, in the course of time, came to build more upon observation, and less upon abstract reasonings. Nicolas Steno, a Dane, published, in 1669, a dissertation *De Solido intra Solidum Naturaliter contento*, in which he says,[6] that though the sides of the hexagonal crystal may vary, *the angles are not changed.* And Dominic Gulielmini, in a *Dissertation on Salts*, published in 1707, says,[7] in a true inductive spirit, 'Nature does not employ all figures, but only certain ones of those which are possible; and of these, the determination is not to be fetched from the brain, or proved à *priori*, but obtained by experiments and observations.' And he speaks[8] with entire decision on this subject: 'Nevertheless since there is here a principle of crystallization, the inclination of the planes and of the angles is always constant.' He even anticipates, very nearly, the views of later crystallographers as to the mode in which crystals are formed from elementary molecules. From this time, many persons laboured and speculated on this subject; as Cappeller, whose *Prodromus Crystallographiæ* appeared at Lucern in 1723; Bourguet, who published *Lettres Philosophiques sur la Formation de Sels et de Cristaux*, at Amsterdam, in 1792; and Henckel, the 'Physicus' of the Elector of Saxony, whose *Pyritologia* came forth in 1725. In this last work we have an example of the description of the various forms of special classes of minerals, (iron pyrites, copper pyrites, and arsenic pyrites;) and an example of the enthusiasm which this apparently dry and laborious study can excite: 'Neither tongue nor stone,' he exclaims,[9] 'can express the satisfaction which I received on setting eyes upon this sinter covered with galena; and thus it constantly happens, that one must have more pleasure in what seems worthless rubbish, than in the purest and most precious ores, if we know aught of minerals.'

Still, however, Henckel[10] disclaims the intention of

---

[6] p. 69.   [7] p. 19.   [8] p. 83.   [9] p. 343.
[10] p. 167.

arranging minerals according to their mathematical forms; and this, which may be considered as the first decided step in the formation of crystallographic mineralogy, appears to have been first attempted by Linnæus. In this attempt, however, he was by no means happy; nor does he himself appear to have been satisfied. He begins his preface by saying, 'Lithology is not what I plume myself upon.' (*Lithologia mihi cristas non eriget.*) Though his sagacity, as a natural historian, led him to see that crystalline form was one of the most definite, and therefore most important, characters of minerals, he failed in profiting by this thought, because, in applying it, he did not employ the light of geometry, but was regulated by what appeared to him resemblances, arbitrarily selected, and often delusive.[11] Thus he derived the form of pyrites from that of vitriol;[12] and brought together alum and diamond on account of their common octohedral form. But he had the great merit of animating to this study one to whom, more perhaps than to any other person, it owes its subsequent progress; I mean Romé de Lisle. 'Instructed,' this writer says, in his preface to his *Essais de Crystallographie,* 'by the works of the celebrated Von Linnée, how greatly the study of the angular form of crystals might become interesting, and fitted to extend the sphere of our mineralogical knowledge, I have followed them in all their metamorphoses with the most scrupulous attention.' The views of Linnæus, as to the importance of this character, had indeed been adopted by several others; as John Hill, the King's gardener at Kew, who, in 1777, published his *Spathogenesia;* and Grignon, who, in 1775, says, ' These crystallizations may give the means of finding a new theory of the generation of crystalline gems.'

The circumstance which threw so much difficulty in the way of those who tried to follow out this thought was, that in consequence of the apparent irregularity of crystals, arising from the extension or contraction of particular sides of the figure, each kind of substance

---

[11] Marx. *Gesch.* p. 97.   [12] *Syst. Nat.* vi. p. 220.

may really appear under many different forms, connected with each other by certain geometrical relations. These may be conceived by considering a certain fundamental form to be cut into new forms in particular ways. Thus if we take a cube, and cut off all the eight corners, till the original faces disappear, we make it an octohedron; and if we stop short of this, we have a figure of fourteen faces, which has been called a *cubo-octohedron*. The first person who appears distinctly to have conceived this *truncation* of angles and edges, and to have introduced the word, is Démeste;[13] although Wallerius[14] had already said, in speaking of the various crystalline forms of calcspar, ' I conceive it would be better not to attend to all differences, lest we be overwhelmed by the number.' And Werner, in his celebrated work *On the External Characters of Minerals*,[15] had formally spoken of *truncation, acuation*, and *acumination*, or replacement by a plane, an edge, a point, respectively, (*abstumpfung, zuschärfung, zuspitzung*,) as ways in which the forms of crystals are modified and often disguised. He applied this process in particular to show the connexion of the various forms which are related to the cube. But still the extension of the process to the whole range of minerals and other crystalline bodies, was due to Romé de Lisle.

---

[13] *Lettres*, 1779, i. 48.

[14] *Systema Mineralogicum*, 1772-5, i. 143.   [15] Leipzig, 1774.

CHAPTER II.

Epoch of Romé de Lisle and Haüy.—Establishment of the Fixity of Crystalline Angles, and
the Simplicity of the Laws of Derivation.

WE have already seen that, before 1780, several
mineralogists had recognized the constancy of
the angles of crystals, and had seen (as Démeste and
Werner,) that the forms were subject to modifications
of a definite kind. But neither of these two thoughts
was so apprehended and so developed, as to supersede
the occasion for a discoverer who should put forward
these principles as what they really were, the materials
of a new and complete science. The merit of this step
belongs jointly to Romé de Lisle and to Haüy. The
former of these two men had already, in 1772, published an *Essai de Crystallographie,* in which he had
described a number of crystals. But in this work his
views are still rude and vague; he does not establish
any connected sequence of transitions in each kind of
substance, and lays little or no stress on the angles.
But in 1783. his ideas[1] had reached a maturity which,
by comparison, excites our admiration. In this he
asserts, in the most distinct manner, the *invariability*
of the angles of crystals of each kind, under all the
changes of relative dimension which the faces may
undergo;[2] and he points out that this invariability
applies only to the *primitive forms,* from each of which
many secondary forms are derived by various changes.[3]
Thus we cannot deny him the merit of having taken
steady hold on both the handles of this discovery,
though something still remained for another to do.
Romé pursues his general ideas into detail with great
labour and skill. He gives drawings of more than
five hundred regular forms; (in his first work he had

---

[1] *Cristallographie, ou Description de Formes propres à tous les
Corps du Règne Minéral.* 3 vols. and 1 vol. of plates.
[2] p. 68.   [3] p. 73.

inserted only one hundred and ten; Linnæus only knew forty;) and assigns them to their proper substances; for instance, thirty to calcspar, and sixteen to felspar. He also invented and used a goniometer. We cannot doubt that he would have been looked upon as a great discoverer, if his fame had not been dimmed by the more brilliant success of his contemporary Haüy.

Réné-Just Haüy is rightly looked upon as the founder of the modern school of crystallography; for all those who have, since him, pursued the study with success, have taken his views for their basis. Besides publishing a system of crystallography and of mineralogy, far more complete than any which had yet appeared, the peculiar steps in the advance which belong to him are, the discovery of the importance of *cleavage*, and the consequent expression of the laws of derivation of secondary from primary forms, by means of the *decrements* of the successive layers of *integrant molecules*.

The latter of these discoveries had already been, in some measure, anticipated by Bergman, who had, in 1773, conceived a hexagonal prism to be built up by the juxta-position of solid rhombs on the planes of a rhombic nucleus.[4] It is not clear[5] whether Haüy was acquainted with Bergman's Memoir, at the time when the cleavage of a hexagonal prism of calcspar, accidentally obtained, led him to the same conception of its structure. But however this might be, he had the indisputable credit of following out this conception with all the vigour of originality, and with the most laborious and persevering earnestness; indeed he made it the business of his life. The hypothesis of a solid, built up of small solids, had this peculiar advantage in reference to crystallography; it rendered a reason of this curious fact;—that a certain series of forms occur in crystals of the same kind, while other forms, apparently intermediate between those which actually

---

[4] *De Formis Crystallorum.* Nov. Act. Reg. Soc. Sc. Ups. 1773.
[5] *Traité de Minér.* 1822, i. 15.

occur, are rigorously excluded. The doctrine of decrements explained this; for by placing a number of regularly-decreasing rows of equal solids, as, for instance, of bricks, upon one another, we might form a regular equal-sided triangle, as the gable of a house; and if the breadth of the gable were one hundred bricks, the height of the triangle might be one hundred, or fifty, or twenty-five; but it would be found that if the height were an intermediate number, as fifty-seven, or forty-three, the edge of the wall would become irregular; and such irregularity is assumed to be inadmissible in the regular structure of crystals. Thus this mode of conceiving crystals allows of certain definite secondary forms, and no others.

The mathematical deduction of the dimensions and proportions of these secondary forms;—the invention of a notation to express them;—the examination of the whole mineral kingdom in accordance with these views;—the production of a work[6] in which they are explained with singular clearness and vivacity;—are services by which Haüy richly earned the admiration which has been bestowed upon him. The wonderful copiousness and variety of the forms and laws to which he was led, thoroughly exercised and nourished the spirit of deduction and calculation which his discoveries excited in him. The reader may form some conception of the extent of his labours, by being told—that the mere geometrical propositions which he found it necessary to premise to his special descriptions, occupy a volume and a half of his work;—that his diagrams are nearly a thousand in number;—that in one single substance (calcspar) he has described forty-seven varieties of form;—and that he has described one kind of crystal (called by him *fer sulfuré parallélique*) which has one hundred and thirty-four faces.

In the course of a long life, he examined, with considerable care, all the forms he could procure of all kinds of mineral; and the interpretation which he gave of the laws of those forms was, in many cases, fixed, by

---

[6] *Traité de Minéralogie*, 1801, 5 vols,

means of a name applied to the mineral in which the form occurred; thus, he introduced such names as *équiaxe, métastatique, unibinaire, perihexahèdre, bisalterne,* and others. It is not now desirable to apply separate names to the different forms of the same mineral species, but these terms answered the purpose, at the time, of making the subjects of study more definite. A symbolical notation is the more convenient mode of designating such forms, and such a notation Haüy invented; but the symbols devised by him had many inconveniences, and have since been superseded by the systems of other crystallographers.

Another of Haüy's leading merits was, as we have already intimated, to have shown, more clearly than his predecessors had done, that the crystalline angles of substances are a criterion of the substances; and that this is peculiarly true of the *angles of cleavage;*— that is, the angles of those edges which are obtained by cleaving a crystal in two different directions;—a mode of division which the structure of many kinds of crystals allowed him to execute in the most complete manner. As an instance of the employment of this criterion, I may mention his separation of the sulphates of baryta and of strontia, which had previously been confounded. Among crystals which in the collections were ranked together as 'heavy spar,' and which were so perfect as to admit of accurate measurement, he found that those which were brought from Sicily, and those of Derbyshire, differed in their cleavage angle by three degrees and a half. ' I could not suppose,' he says,[7] ' that this difference was the effect of any law of decrement; for it would have been necessary to suppose so rapid and complex a law, that such an hypothesis might have been justly regarded as an abuse of the theory.' He was, therefore, in great perplexity. But a little while previous to this, Klaproth had discovered that there is an earth which, though in many respects it resembles baryta, is different from it in other respects; and this earth, from the place where it was found (in

---

[7] *Traité,* ii. 320.

Scotland), had been named *Strontia*. The French chemists had ascertained that the two earths had, in some
cases, been mixed or confounded; and Vauquelin, on
examining the Sicilian crystals, found that their base
was strontia, and not, as in the Derbyshire ones, baryta.
The riddle was now read; all the crystals with the
larger angle belong to the one, all those with the
smaller, to the other, of these two sulphates; and
crystallometry was clearly recognized as an authorized
test of the difference of substances which nearly resemble
each other.

Enough has been said, probably, to enable the reader
to judge how much each of the two persons, now under
review, contributed to crystallography. It would be
unwise to compare such contributions to science with
the great discoveries of astronomy and chemistry; and
we have seen how nearly the predecessors of Romé
and Haüy had reached the point of knowledge on which
these two crystallographers took their stand. But yet
it is impossible not to allow, that in these discoveries,
which thus gave form and substance to the science of
crystallography, we have a manifestation of no common
sagacity and skill. Here, as in other discoveries, were
required ideas and facts;—clearness of geometrical conception which could deal with the most complex relations of form; a minute and extensive acquaintance
with actual crystals; and the talent and habit of referring
these facts to the general ideas. Haüy, in particular,
was happily endowed for his task. Without being a
great mathematician, he was sufficiently a geometer to
solve all the problems which his undertaking demanded;
and though the mathematical reasoning might have
been made more compendious by one who was more
at home in mathematical generalization, probably this
could hardly have been done without making the subject less accessible and less attractive to persons mode-
rately disciplined in mathematics. In all his reasonings
upon particular cases, Haüy is acute and clear; while
his general views appear to be suggested rather by a
lively fancy than by a sage inductive spirit: and though
he thus misses the character of a great philosopher, the

vivacity of style, and felicity and happiness of illustration, which grace his book, and which agree well with the character of an Abbé of the old French monarchy, had a great and useful influence on the progress of the subject.

Unfortunately Romé de Lisle and Haüy were not only rivals, but in some measure enemies. The former might naturally feel some vexation at finding himself, in his later years (he died in 1790), thrown into shade by his more brilliant successor. In reference to Haüy's use of cleavage, he speaks[8] of 'innovators in crystallography, who may properly be called *crystalloclasts.*' Yet he adopted, in great measure, the same views of the formation of crystals by laminæ,[9] which Haüy illustrated by the destructive process at which he thus sneers. His sensitiveness was kept alive by the conduct of the Academy of Sciences, which took no notice of him and his labours;[10] probably because it was led by Buffon, who disliked Linnæus, and might dislike Romé as his follower; and who, as we have seen, despised crystallography. Haüy revenged himself by rarely mentioning Romé in his works, though it was manifest that his obligations to him were immense; and by recording his errours while he corrected them. More fortunate than his rival, Haüy was, from the first, received with favour and applause. His lectures at Paris were eagerly listened to by persons from all quarters of the world. His views were, in this manner, speedily diffused; and the subject was soon pursued, in various ways, by mathematicians and mineralogists in every country of Europe.

---

[8] Pref., p. xxvii.   [9] T. ii. p. 21.
[10] Marx. *Gesch. d. Kryst.* 130.

CHAPTER III.

RECEPTION AND CORRECTIONS OF THE HAUÏAN CRYSTALLOGRAPHY.

I HAVE not hitherto noticed the imperfections of the crystallographic views and methods of Haüy, because my business in the last section was to mark the permanent additions he made to the science. His system did, however, require completion and rectification in various points; and in speaking of the crystallographers of the subsequent time, who may all be considered as the cultivators of the Haüïan doctrines, we must also consider what they did in correcting them.

The three main points in which this improvement was needed were;—a better determination of the crystalline forms of the special substances;—a more general and less arbitrary method of considering crystalline forms according to their symmetry;—and a detection of more general conditions by which the crystalline angle is regulated. The first of these processes may be considered as the natural sequel of the Haüïan epoch: the other two must be treated as separate steps of discovery.

When it appeared that the angle of natural or of cleavage faces could be used to determine the differences of minerals, it became important to measure this angle with accuracy. Haüy's measurements were found very inaccurate by many succeeding crystallographers; Mohs says[1] that they are so generally inaccurate, that no confidence can be placed in them. This was said, of course, according to the more rigorous notions of accuracy to which the establishment of Haüy's system led. Among the persons who principally laboured in ascertaining, with precision, the crystalline angles of minerals, were several Englishmen, especially Wollaston, Phillips, and Brooke. Wollaston, by the invention of

---

[1] Marx. p. 153.

his Reflecting Goniometer, placed an entirely new degree of accuracy within the reach of the crystallographer; the angle of two faces being, in this instrument, measured by means of the reflected images of bright objects seen in them, so that the measure is the more accurate the more minute the faces are. In the use of this instrument, no one was more laborious and successful than William Phillips, whose power of apprehending the most complex forms with steadiness and clearness, led Wollaston to say that he had 'a geometrical sense.' Phillips published a Treatise on Mineralogy, containing a great collection of such determinations; and Mr. Brooke, a crystallographer of the same exact and careful school, has also published several works of the same kind. The precise measurement of crystalline angles must be the familiar employment of all who study crystallography; and, therefore, any further enumeration of those who have added, in this way, to the stock of knowledge, would be superfluous.

Nor need I dwell long on those who added to the knowledge which Haüy left, of derived forms. The most remarkable work of this kind was that of Count Bournon, who published a work on a single mineral (calcspar) in three quarto volumes.[2] He has here given representations of seven hundred forms of crystals, of which, however, only fifty-six are essentially different. From this example the reader may judge what a length of time, and what a number of observers and calculators, were requisite to exhaust the subject.

If the calculations, thus occasioned, had been conducted upon the basis of Haüy's system, without any further generalization, they would have belonged to that process, the natural sequel of inductive discoveries, which we call *deduction;* and would have needed only a very brief notice here. But some additional steps were made in the upward road to scientific truth, and of these we must now give an account.

---

[2] *Traité complet de la Chaux Carbonatée et d'Aragonite,* par M. le Comte de Bournon. London, 1808.

CHAPTER IV.

ESTABLISHMENT OF THE DISTINCTION OF SYSTEMS OF CRYSTALLIZATION.—WEISS AND MOHS.

IN Haüy's views, as generally happens in new systems, however true, there was involved something that was arbitrary, something that was false or doubtful, something that was unnecessarily limited. The principal points of this kind were;—his having made the laws of crystalline derivation depend so much upon cleavage;—his having assumed an atomic constitution of bodies as an essential part of his system;—and his having taken a set of primary forms, which, being selected by no general view, were partly superfluous, and partly defective.

How far evidence, such as has been referred to by various philosophers, has proved, or can prove, that bodies are constituted of indivisible atoms, will be more fully examined in the work which treats of the Philosophy of this subject. There can be little doubt that the portion of Haüy's doctrine which most riveted popular attention and applause, was his dissection of crystals, in a manner which was supposed to lead actually to their ultimate material elements. Yet it is clear, that since the solids given by cleavage are, in many cases, such as cannot make up a solid space, the primary conception, of a necessary geometrical identity between the results of division and the elements of composition, which is the sole foundation of the supposition that crystallography points out the actual elements, disappears on being scrutinized: and when Haüy, pressed by this difficulty, as in the case of fluor-spar, put his integrant octohedral molecules together, touching by the edges only, his method became an empty geometrical diagram, with no physical meaning.

The real fact, divested of the hypothesis which was contained in the fiction of decrements, was, that when the relation of the derivative to the primary faces is expressed by means of numerical indices, these numbers are integers, and generally very small ones; and this was the form which the law gradually assumed, as the method of derivation was made more general and simple by Weiss and others.

'When, in 1809, I published my Dissertation,' says Weiss,[1] 'I shared the common opinion as to the necessity of the assumption and the reality of the existence of a primitive form, at least in a sense not very different from the usual sense of the expression. While I sought,' he adds, referring to certain doctrines of general philosophy which he and others entertained, 'a *dynamical* ground for this, instead of the untenable atomistic view, I found that, out of my primitive forms, there was gradually unfolded to my hands, that which really governs them, and is not affected by their casual fluctuations, the fundamental relations of those Dimensions according to which a multiplicity of internal oppositions, necessarily and mutually interdependent, are developed in the mass, each having its own polarity; so that the crystalline character is coextensive with these polarities.'

The 'Dimensions' of which Weiss here speaks, are the *Axes of Symmetry* of the crystal; that is, those lines, in reference to which, every face is accompanied by other faces, having like positions and properties. Thus a rhomb, or more properly a *rhombohedron*,[2] of calc-spar may be placed with one of its obtuse corners uppermost, so that all the three faces which meet there are equally inclined to the vertical line. In this position, every derivative face, which is obtained by any modification of the faces or edges of the rhombohedron, implies either three or six such derivative faces; for no one of the three upper faces of the rhombohedron has any character or property different from the other two; and, therefore, there is no reason for the existence of a derivative from one of these primitive faces, which does not equally hold for the other primitive

---

[1] *Mem. Acad. Berl.* 1816, p. 307.

[2] I use this name for the solid figure, since *rhomb* has always been used for a plane figure.

faces. Hence the derivative forms will, in all cases, contain none but faces connected by this kind of correspondence. The axis thus made vertical will be an Axis of Symmetry, and the crystal will consist of three divisions, ranged round this axis, and exactly resembling each other. According to Weiss's nomenclature, such a crystal is 'three-and-three-membered.'

But this is only one of the kinds of symmetry which crystalline forms may exhibit. They may have *three axes* of complete and *equal* symmetry at right angles to each other, as the cube and the regular octohedron; —or, *two axes* of equal symmetry, perpendicular to each other and to a *third axis*, which is not affected with the same symmetry with which they are; such a figure is a square pyramid;—or they may have *three* rectangular *axes*, all of *unequal* symmetry, the modifications. referring to each axis separately from the other two.

These are essential and necessary distinctions of crystalline form; and the introduction of a classification of forms founded on such relations, or as they were called, *Systems of Crystallization*, was a great improvement upon the divisions of the earlier crystallographers, for those divisions were separated according to certain arbitrarily-assumed primary forms. Thus Romé de Lisle's fundamental forms were, the tetrahedron, the cube, the octohedron, the rhombic prism, the rhombic octohedron, the dodecahedron with triangular faces: Haüy's primary forms are the cube, the rhombohedron, the oblique rhombic prism, the right rhombic prism, the rhombic dodecahedron, the regular octohedron, tetrahedron, and six-sided prism, and the bipyramidal dodecahedron. This division, as I have already said, errs both by excess and defect, for some of these primary forms might be made derivatives from others; and no solid reason could be assigned why they were not. Thus the cube may be derived from the tetrahedron, by truncating the edges; and the rhombic dodecahedron again from the cube, by truncating its edges; while the square pyramid could not be legitimately identified with the derivative of any of these forms; for if we were to derive it from the rhombic prism, why should the acute angles always suffer decrements corresponding in a certain way to those of the obtuse angles, as they must do in order to give rise to a square pyramid?

The introduction of the method of reference to Systems of Crystallization has been a subject of controversy, some ascribing this valuable step to Weiss, and some to Mohs.[3] It appears, I think, on the whole, that Weiss first published works in which the method is employed; but that Mohs, by applying it to all the known species of minerals, has had the merit of making it the basis of real crystallography. Weiss, in 1809, published a Dissertation *On the mode of investigating the principal geometrical character of crystalline forms*, in which he says,[4] 'No part, line, or quantity, is so important as the axis; no consideration is more essential or of a higher order than the relation of a crystalline plane to the axis;' and again, 'An axis is any line governing the figure, about which all parts are similarly disposed, and with reference to which they correspond mutually.' This he soon followed out by examination of some difficult cases, as Felspar and Epidote. In the Memoirs of the Berlin Academy,[5] for 1814-5, he published *An Exhibition of the natural Divisions of Systems of Crystallization.* In this Memoir, his divisions are as follows:—The *regular* system, the *four-membered,* the *two-and-two-membered,* the *three-and-three-membered,* and some others of inferior degrees of symmetry. These divisions are by Mohs (*Outlines of Mineralogy,* 1822,) termed the *tessular, pyramidal, prismatic,* and *rhombohedral* systems respectively. Hausmann, in his *Investigations concerning the Forms of Inanimate Nature,*[6] makes a nearly corresponding arrangement;—the *isometric, monodimetric, trimetric,* and *monotrimetric;* and one or other of these sets of terms have been adopted by most succeeding writers.

In order to make the distinctions more apparent, I

---

[3] *Edin. Phil. Trans.* 1823, vols. xv. and xvi. [4] pp. 16, 42.
[5] Ibid. [6] Göttingen, 1821.

have purposely omitted to speak of the systems which arise when the *prismatic* system loses some part of its symmetry;—when it has only half or a quarter its complete number of faces;—or, according to Mohs's phraseology, when it is *hemihedral* or *tetartohedral.* Such systems are represented by the singly-oblique or doubly-oblique prism; they are termed by Weiss *two-and-one membered,* and *one-and-one-membered;* by other writers, *Monoklinometric,* and *Triklinometric* Systems. There are also other peculiarities of Symmetry, such, for instance, as that of the *plagihedral* faces of quartz, and other minerals.

The introduction of an arrangement of crystalline forms into systems, according to their degree of symmetry, was a step which was rather founded on a distinct and comprehensive perception of mathematical relations, than on an acquaintance with experimental facts, beyond what earlier mineralogists had possessed. This arrangement was, however, remarkably confirmed by some of the properties of minerals which attracted notice about the time now spoken of, as we shall see in the next chapter.

CHAPTER V.

DIFFUSION of the Distinction of Systems.—The distinction of systems of crystallization was so far founded on obviously true views, that it was speedily adopted by most mineralogists. I need not dwell on the steps by which this took place. Mr. Haidinger's translation of Mohs was a principal occasion of its introduction in England. As an indication of dates, bearing on this subject, perhaps I may be allowed to notice, that there appeared in the *Philosophical Transactions for* 1825, *A General Method of Calculating the Angles of Crystals*, which I had written, and in which I referred only to Haüy's views; but that in 1826,[1] I published a Memoir *On the Classification of Crystalline Combinations*, founded on the methods of Weiss and Mohs, especially the latter; with which I had in the mean time become acquainted, and which appeared to me to contain their own evidence and recommendation. General methods, such as was attempted in the Memoir just quoted, are part of that process in the history of sciences, by which, when the principles are once established, the mathematical operation of deducing their consequences is made more and more general and symmetrical: which we have seen already exemplified in the history of celestial mechanics after the time of Newton. It does not enter into our plan, to dwell upon the various steps in this way made by Levy, Naumann, Grassmann, Kupffer, Hessel, and by Professor Miller among ourselves. I may notice that one great improvement was, the method introduced by Monteiro and Levy, of determining the laws of derivation of forces by means of the *parallelisms of edges;* which was afterwards

---

[1] *Camb. Trans.* vol. ii. p. 391.

extended so that faces were considered as belonging to *zones*. Nor need I attempt to enumerate (what indeed it would be difficult to describe in words) the various methods of *notation* by which it has been proposed to represent the faces of crystals, and to facilitate the calculations which have reference to them.

[2nd Ed.] [My Memoir of 1825 depended on the views of Haüy in so far as that I started from his 'primitive forms;' but, being a general method of expressing all forms by co-ordinates, it was very little governed by those views. The mode of representing crystalline forms which I proposed seemed to contain its own evidence of being more true to nature than Haüy's theory of decrements, inasmuch as my method expressed the faces by much lower numbers. I determine a face by means of the dimensions of the primary form *divided* by certain numbers; Haüy had expressed the face virtually by the same dimensions *multiplied* by numbers. In cases where my notation gives such numbers as (3, 4, 1), (1, 3, 7), (5, 1, 19), his method involves the higher numbers (4, 3, 12), (21, 7, 3), (19, 95, 5). My method however has, I believe, little value as a method of '*calculating* the angles of crystals.'

M. Neumann, of Königsberg, introduced a very convenient and elegant method of representing the position of faces of crystals by corresponding points on the surface of a circumscribing sphere. He gave (in 1823) the laws of the derivation of crystalline faces, expressed geometrically by the intersection of zones, (*Beiträge zur Krystallonomie.*) The same method of indicating the position of faces of crystals was afterwards, together with the notation, re-invented by M. Grassmann, (*Zur Krystallonomie und Geometrischen Combinationslehre*, 1829.) Aiding himself by the suggestions of these writers, and partly adopting my method, Prof. Miller has produced a work on Crystallography remarkable for mathematical elegance and symmetry; and has given expressions really useful for calculating the angles of crystalline faces, (*A Treatise on Crystallography.* Cambridge, 1839.)]

Confirmation of the Distinction of Systems by the Optical Properties of Minerals.—Brewster.—I must not omit to notice the striking confirmation which the distinction of systems of crystallization received from optical discoveries, especially those of Sir D. Brewster. Of the history of this very rich and beautiful department of science, we have already given some account, in speaking of Optics. The first facts which were noticed, those relating to double refraction, belonged exclusively to crystals of the rhombohedral system. The splendid phenomena of the rings and lemniscates produced by dipolarizing crystals, were afterwards discovered; and these were, in 1817, classified by Sir David Brewster, according to the crystalline forms to which they belong. This classification, on comparison with the distinction of Systems of Crystallization, resolved itself into a necessary relation of mathematical symmetry: all crystals of the pyramidal and rhombohedral systems, which from their geometrical character have a single axis of symmetry, are also optically uniaxal, and produce by dipolarization circular rings; while the prismatic system, which has no such single axis, but three unequal axes of symmetry, is optically biaxal, gives lemniscates by dipolarized light, and, according to Fresnel's theory, has three rectangular axes of unequal elasticity.

[2nd Ed.] [I have placed Sir David Brewster's arrangement of crystalline forms in this chapter, as an event belonging to the *confirmation* of the distinctions of forms introduced by Weiss and Mohs; because that arrangement was established, not on crystallographical, but on optical grounds. But Sir David Brewster's optical discovery was a much greater step in science than the systems of the two German crystallographers; and even in respect to the crystallographical principle, Sir D. Brewster had an independent share in the discovery. He divided crystalline forms into three classes, enumerating the Haüian 'primitive forms' which belonged to each; and as he found some exceptions to this classification, (such as idocrase, &c.,) he ventured to pronounce that in those substances the received primitive forms were probably erroneous; a judgment which was soon confirmed by a closer crystallographical scrutiny. He also showed his perception of the mineralogical importance of his discovery by publishing it, not only in the *Phil. Trans.* (1818), but also in the *Transactions of the Wernerian Society of Natural History.* In a second paper inserted in this latter series, read in 1820, he further notices Mohs's System of Crystallography, which had then recently appeared, and points out its agreement with his own.

Another reason why I do not make this great optical discovery a cardinal point in the history of crystallography is, that as a crystallographical system it is incomplete. Although we are thus led to distinguish the *tessular* and the *prismatic* systems (using Mohs's terms) from the *rhombohedral* and the *square prismatic*, we are not led to distinguish the latter two from each other; inasmuch as they have no optical difference of character. But this distinction is quite essential in crystallography; for these two systems have faces formed by laws as different as those of the other two systems.

Moreover, Weiss and Mohs not only divided crystalline forms into certain classes, but showed that by doing this, the derivation of all the existing forms from the fundamental ones assumed a new aspect of simplicity and generality; and this was the essential part of what they did.

On the other hand, I do not think it is too much to say, as I have elsewhere said,[2] that 'Sir D. Brewster's optical experiments must have led to a classification of crystals into the above systems, or something nearly equivalent, even if crystals had not been so arranged by attention to their forms.']

Many other most curious trains of research have confirmed the general truth, that the degree and kind of geometrical symmetry corresponds exactly with the symmetry of the optical properties. As an instance

---

[2] *Philosophy of the Inductive Sciences*, B. viii. C. iii. Art. 3.

of this, eminently striking for its singularity, we may notice the discovery of Sir John Herschel, that the *plagihedral* crystallization of quartz, by which it exhibits faces *twisted* to the right or the left, is accompanied by right-handed or left-handed circular polarization respectively. No one acquainted with the subject can now doubt, that the correspondence of geometrical and optical symmetry is of the most complete and fundamental kind.

[2nd Ed.] [Our knowledge with respect to the positions of the optical axes of oblique prismatic crystals is still imperfect. It appears to be ascertained that, in singly oblique crystals, one of the axes of optical elasticity coincides with the rectangular crystallographic axis. In doubly-oblique crystals, one of the axes of optical elasticity is, in many cases, coincident with the axis of a principal zone. I believe no more determinate laws have been discovered.]

Thus the highest generalizations at which mathematical crystallographers have yet arrived, may be considered as fully established; and the science of Crystallography, in the condition in which these place it, is fit to be employed as one of the members of Mineralogy, and thus to fill its appropriate place and office.

CHAPTER VI.

DISCOVERY of Isomorphism. Mitscherlich.—The discovery of which we now have to speak may appear at first sight too large to be included in the history of crystallography, and may seem to belong rather to chemistry. But it is to be recollected that crystallography, from the time of its first assuming importance in the hands of Haüy, founded its claim to notice entirely upon its connexion with chemistry; crystalline forms were properties of *something;* but *what* that something was, and how it might be modified without becoming something else, no crystallographer could venture to decide, without the aid of chemical analysis. Haüy had assumed, as the general result of his researches, that the same chemical elements, combined in the same proportions, would always exhibit the same crystalline form; and reciprocally, that the same form and angles (except in the obvious case of the tessular system, in which the angles are determined by its *being* the tessular system,) implied the same chemical constitution. But this dogma could only be considered as an approximate conjecture; for there were many glaring and unexplained exceptions to it. The explanation of several of these was beautifully described by the discovery that there are various elements which are *isomorphous* to each other; that is, such that one may take the place of another without altering the crystalline form; and thus the chemical composition may be much changed, while the crystallographic character is undisturbed.

This truth had been caught sight of, probably as a guess only, by Fuchs as early as 1815. In speaking of a mineral which had been called Gehlenite, he says,

'I hold the oxide of iron, not for an essential component part of this genus, but only as a *vicarious* element,
replacing so much lime. We shall find it necessary
to consider the results of several analyses of mineral
bodies in this point of view, if we wish, on the one
hand, to bring them into agreement with the doctrine
of chemical proportions, and on the other, to avoid
unnecessarily splitting up genera.' In a lecture *On
the Mutual Influence of Chemistry and Mineralogy*,[1] he
again draws attention to his term *vicarious* (*vicari-
rende*,) which undoubtedly expresses the nature of the
general law afterwards established by Mitscherlich in
1822.

But Fuchs's conjectural expression was only a
prelude to Mitscherlich's experimental discovery of
isomorphism. Till many careful analyses had given
substance and signification to this conception of
vicarious elements, it was of small value. Perhaps no
one was more capable than Berzelius of turning to the
best advantage any ideas which were current in the
chemical world; yet we find him,[2] in 1820, dwelling
upon a certain vague view of these cases,—that ' oxides
which contain equal doses of oxygen must have their
general properties common ;' without tracing it to any
definite conclusions. But his scholar, Mitscherlich,
gave this proposition a real crystallographical import.
Thus he found that the carbonates of lime (calc-spar,)
of magnesia, of protoxide of iron, and of protoxide of
manganese, agree in many respects of form, while the
homologous angles vary through one or two degrees
only; so again the carbonates of baryta, strontia, lead,
and lime (arragonite), agree nearly; the different kinds
of felspar vary only by the substitution of one alkali
for another; the phosphates are almost identical with
the arseniates of several bases. These, and similar
results, were expressed by saying that, in such cases,
the bases, lime, protoxide of iron, and the rest, are

---

[1] Munich, 1820.

[2] *Essay on the Theory of Chemical Proportions*, p. 122.

isomorphous; or in the latter instance, that the arsenic and phosphoric acids are isomorphous.

Since, in some of these cases, the substitution of one element of the isomorphous group for another does alter the angle, though slightly, it has since been proposed to call such groups *plesiomorphous.*

This discovery of isomorphism was of great importance, and excited much attention among the chemists of Europe. The history of its reception, however, belongs, in part, to the classification of minerals; for its effect was immediately to metamorphose the existing chemical systems of arrangement. But even those crystallographers and chemists who cared little for general systems of classification, received a powerful impulse by the expectation, which was now excited, of discovering definite laws connecting chemical constitution with crystalline form. Such investigations were soon carried on with great activity. Thus, at a recent period, Abich analysed a number of tessular minerals, spinelle, pleonaste, gahnite, franklinite, and chromic iron oxide; and seems to have had some success in giving a common type to their chemical formulæ, as there is a common type in their crystallization.

[2nd Ed.] [It will be seen by the above account that Prof. Mitscherlich's merit in the great discovery of Isomorphism is not at all narrowed by the previous conjectures of M. Fuchs. I am informed, moreover, that M. Fuchs afterwards (in Schweigger's *Journal*) retracted the opinions he had put forwards on this subject.]

Dimorphism.—My business is, to point out the connected truths which have been obtained by philosophers, rather than insulated difficulties which still stand out to perplex them. I need not, therefore, dwell on the curious cases of *dimorphism;* cases in which the same definite chemical compound of the same elements appears to have two different forms; thus the carbonate of lime has two forms, *calc-spar* and *arragonite,* which belong to different systems of crystallization. Such facts may puzzle us; but they hardly interfere with any received general truths, because we have as yet no truths of very high order respecting the connexion of chemical constitution and crystalline form. Dimorphism does not interfere with isomorphism; the two classes of facts stand at the same stage of inductive generalization, and we wait for some higher truth, which shall include both, and rise above them.

[2nd Ed.] [For additions to our knowledge of the Dimorphism of Bodies, see Professor Johnstone's valuable *Report* on that subject in the *Reports of the British Association* for 1837. Substances have also been found which are *trimorphous*. We owe to Professor Mitscherlich the discovery of dimorphism, as well as of isomorphism : and to him also we owe the greater part of the knowledge to which these discoveries have led.]

CHAPTER VII.

ATTEMPTS TO ESTABLISH THE FIXITY OF OTHER PHYSICAL PROPERTIES.—WERNER.

THE reflections from which it appeared, (at the end of the last Book,) that in order to obtain general knowledge respecting bodies, we must give scientific fixity to our appreciation of their properties, applies to their other properties as well as to their crystalline form. And though none of the other properties have yet been referred to standards so definite as that which geometry supplies for crystals, a system has been introduced which makes their measures far more constant and precise than they are to a common undisciplined sense.

The author of this system was Abraham Gottlob Werner, who had been educated in the institutions which the elector of Saxony had established at the mines of Freiberg. Of an exact and methodical intellect, and of great acuteness of the senses, Werner was well fitted for the task of giving fixity to the appreciation of outward impressions; and this he attempted in his *Dissertation on the External Characters of Fossils,* which was published at Leipzig in 1774. Of the precision of his estimation of such characters, we may judge from the following story, told by his biographer Frisch.[1] One of his companions had received a quantity of pieces of amber, and was relating to Werner, then very young, that he had found in the lot one piece from which he could extract no signs of electricity. Werner requested to be allowed to put his hand in the bag which contained these pieces, and immediately drew out the unelectrical piece. It was yellow chalcedony, which is distinguishable from amber by its weight and coldness.

The principal external characters which were sub-

---

[1] *Werner's Leben*, p. 26.

jected by Werner to a systematic examination, were colour, lustre, hardness, and specific gravity. His subdivisions of the first character (*Colour*,) were very numerous; yet it cannot be doubted that if we recollect them by the eye, and not by their names, they are definite and valuable characters, and especially the metallic colours. Breithaupt, merely by the aid of this character, distinguished two new compounds among the small grains found along with the grains of platinum, and usually confounded with them. The kinds of *Lustre*, namely, *glassy*, *fatty*, *adamantine*, *metallic*, are, when used in the same manner, equally valuable. *Specific Gravity* obviously admits of a numerical measure; and the *Hardness* of a mineral was pretty exactly defined by the substances which it would scratch, and by which it was capable of being scratched.

Werner soon acquired a reputation as a mineralogist, which drew persons from every part of Europe to Freiberg in order to hear his lectures; and thus diffused very widely his mode of employing external characters. It was, indeed, impossible to attend so closely to these characters as the Wernerian method required, without finding that they were more distinctive than might at first sight be imagined; and the analogy which this mode of studying Mineralogy established between that and other branches of Natural History, recommended the method to those in whom a general inclination to such studies was excited. Thus Professor Jameson of Edinburgh, who had been one of the pupils of Werner at Freiberg, not only published works in which he promulgated the mineralogical doctrines of his master, but established in Edinburgh a 'Wernerian Society,' having for its object the general cultivation of Natural History.

Werner's standards and nomenclature of external characters were somewhat modified by Mohs, who, with the same kind of talents and views, succeeded him at Freiberg. Mohs reduced hardness to numerical measure by selecting ten known minerals, each harder than the other in order, from *talc* to *corundum* and *diamond*, and by making the place which these minerals occupy in the list, the numerical measure of the hardness of those which are compared with them. The result of the application of this fixed measurement and nomenclature of external characters will appear in the History of Classification, to which we now proceed.

CHAPTER IV.

ATTEMPTS TO DISCOVER GENERAL LAWS IN GEOLOGY.

Sect. I.—*General Geological Phenomena.*

BESIDES thus noticing such features in the rocks of each country as were necessary to the identification of the strata, geologists have had many other phenomena of the earth's surface and materials presented to their notice; and these they have, to a certain extent, attempted to generalize, so as to obtain on this subject what we have elsewhere termed the Laws of Phenomena, which are the best materials for physical theory. Without dwelling long upon these, we may briefly note some of the most obvious. Thus it has been observed that mountain ranges often consist of a ridge of subjacent rock, on which lie, on each side, strata sloping from the ridge. Such a ridge is an *Anticlinal Line*, a *Mineralogical Axis*. The sloping strata present their *Escarpments*, or steep edges, to this axis. Again, in mining countries, the *Veins* which contain the ore are usually a system of *parallel* and nearly vertical partitions in the rock; and these are, in very many cases, intersected by another system of veins parallel to each other, and nearly *perpendicular* to the former. Rocky regions are often intersected by *Faults*, or fissures interrupting the strata, in which the rock on one side the fissure appears to have been at first continuous with that on the other, and shoved aside or up or down after the fracture. Again, besides these larger fractures, rocks have *Joints*,—separations, or tendencies to separate in some directions rather than in others; and a *slaty Cleavage*, in which the parallel subdivisions may be carried on, so as to produce laminæ of indefinite thinness. As an example of those

laws of phenomena of which we have spoken, we may
instance the general law asserted by Prof. Sedgwick,
(not, however, as free from exception,) that in one
particular class of rocks the slaty Cleavage *never* coincides with the Direction of the strata.

The phenomena of metalliferous veins may be
referred to, as another large class of facts which
demand the notice of the geologist. It would be
difficult to point out briefly any general laws which
prevail in such cases; but in order to show the curious
and complex nature of the facts, it may be sufficient
to refer to the description of the metallic veins of
Cornwall by Mr. Carne;[1] in which the author maintains that their various contents, and the manner in
which they cut across, and *stop*, or *shift*, each other,
leads naturally to the assumption of veins of no less
than six or eight different ages in one kind of rock.

Again, as important characters belonging to the
physical history of the earth, and therefore to geology,
we may notice all the general laws which refer to its
temperature;—both the laws of climate, as determined
by the *isothermal lines*, which Humboldt has drawn,
by the aid of very numerous observations made in all
parts of the world; and also those still more curious
facts, of the increase of temperature which takes place
as we descend in the solid mass. The latter circumstance, after being for a while rejected as a fable, or
explained away as an accident, is now generally
acknowledged to be the true state of things in many
distant parts of the globe, and probably in all.

Again, to turn to cases of another kind: some
writers have endeavoured to state in a general manner
laws according to which the members of the geological
series succeed each other; and to reduce apparent
anomalies to order of a wider kind. Among those
who have written with such views, we may notice
Alexander von Humboldt, always, and in all sciences,
foremost in the race of generalization. In his attempt
to extend the doctrine of geological equivalents from

---

[1] *Transactions of the Geol. Soc. of Cornwall*, vol. ii.

the rocks of Europe[2] to those of the Andes, he has marked by appropriate terms the general modes of geological succession. 'I have insisted,' he says,[3] 'principally upon the phenomena of *alternation, oscillation,* and *local suppression,* and on those presented by the *passages* of formations from one to another, by the effect of an *interior developement.*'

The phenomena of alternation to which M. de Humboldt here refers are, in fact, very curious: as exhibiting a mode in which the transitions from one formation to another may become gradual and insensible, instead of sudden and abrupt. Thus the coal measures in the south of England are above the mountain limestone; and the distinction of the formations is of the most marked kind. But as we advance northward into the coal-field of Yorkshire and Durham, the subjacent limestone begins to be subdivided by thick masses of sandstone and carbonaceous strata, and passes into a complex deposit, not distinguishable from the overlying coal measures; and in this manner the transition from the limestone to the coal is made by alternation. Thus, to use another expression of M. de Humboldt's, in ascending from the limestone, the coal, before we quit the subjacent stratum, *preludes* to its fuller exhibition in the superior beds.

Again, as to another point: geologists have gone on up to the present time endeavouring to discover general laws and facts, with regard to the position of mountain and mineral masses upon the surface of the earth. Thus M. Von Buch, in his physical description of the Canaries, has given a masterly description of the lines of volcanic action and volcanic products, all over the globe. And, more recently, M. Elie de Beaumont has offered some generalizations of a still wider kind. In this new doctrine, those mountain ranges, even in distant parts of the world, which are of the same age, according to the classifications already

---

[2] *Gissement des Roches dans les deux Hémisphères,* 1823.
[3] Pref. p vi.

spoken of, are asserted to be parallel[4] to each other, while those ranges which are of different ages lie in different directions. This very wide and striking proposition may be considered as being at present upon its trial among the geologists of Europe.[5]

Among the organic phenomena, also, which have been the subject of geological study, general laws of a very wide and comprehensive kind have been suggested, and in a greater or less degree confirmed by adequate assemblages of facts. Thus M. Adolphe Brongniart has not only, in his *Fossil Flora*, represented and skilfully restored a vast number of the plants of the ancient world; but he has also, in the *Prodromus* of the work, presented various important and striking views of the general character of the vegetation of former periods, as insular or continental, tropical or temperate. And M. Agassiz, by the examination of an incredible number of specimens and collections of fossil fish, has been led to results which, expressed in terms of his own ichthyological classification, form remarkable general laws. Thus, according to him,[6] when we go below the lias, we lose all traces of two of the four orders under which he comprehends all known kinds of fish; namely, the *Cycloïdean* and the *Ctenoïdean*; while the other two orders, the *Ganoïdean* and *Placoïdean*, rare in our days, suddenly appear in great numbers, together with large sauroid and carnivorous fishes. Cuvier, in constructing his great work on ichthyology, transferred to M. Agassiz the whole subject of fossil fishes, thus showing how highly he esteemed

---

[4] We may observe that the notion of parallelism, when applied to lines drawn on *remote* portions of a globular surface, requires to be interpreted in so arbitrary a manner, that we can hardly imagine it to express a physical law.

[5] Mr. Lyell, in the sixth edition of his *Principles*, B. i. c. xii., has combated the hypothesis of M. Elie de Beaumont, stated in the text. He has argued both against the catastrophic character of the elevation of mountain chains, and the parallelism of the contemporaneous ridges. It is evident that the former doctrine may be true, though the latter be shown to be false.

[6] Greenough, *Address to Geol. Soc.* 1835, p. 19.

his talents as a naturalist. And M. Agassiz has shown himself worthy of his great predecessor in geological natural history, not only by his acuteness and activity, but by the comprehensive character of his zoological philosophy, and by the courage with which he has addressed himself to the vast labours which lie before him. In his *Report on the Fossil Fish discovered in England*, published in 1835, he briefly sketches some of the large questions which his researches have suggested; and then adds,[7] 'Such is the meagre outline of a history of the highest interest, full of curious episodes, but most difficult to relate. To unfold the details which it contains will be the business of my life.'

[2nd Ed.] [In proceeding downwards through the series of formations into which geologists have distributed the rocks of the earth, one class of organic forms after another is found to disappear. In the Tertiary Period we find all the classes of the present world: Mammals, Birds, Reptiles, Fishes, Crustaceans, Mollusks, Zoophytes. In the Secondary Period, from the Chalk down to the New Red Sandstone, Mammals are not found, with the minute exception of the marsupial *amphitherium* and *phascolotherium* in the Stonesfield slate. In the Carboniferous and Devonian period we have no large Reptiles, with, again, a minute amount of exception. In the lower part of the Silurian rocks, Fishes vanish, and we have no animal forms but Mollusks, Crustaceans and Zoophytes.

The Carboniferous, Devonian and Silurian formations, thus containing the oldest forms of life, have been termed *palæozoic*. The boundaries of the life-bearing series have not yet been determined; but the series has in which vertebrated animals do not appear been provisionally termed *protozoic*, and the lower Silurian rocks may probably be looked upon as its upper members. Below this, geologists place a *hypozoic* or *azoic* series of rocks.

Geologists differ as to the question whether these changes in the inhabitants of the globe were made

---

[7] *Brit. Assoc. Report*, p. 72.

by determinate steps or by insensible gradations. M. Agassiz has been led to the conviction that the organized population of the globe was renewed in the interval of each principal member of its formations.[8] Mr. Lyell, on the other hand, conceives that the change in the collection of organized beings was gradual, and has proposed on this subject an hypothesis which I shall hereafter consider.]

Sect. 2.—Transition to Geological Dynamics.

WHILE we have been giving this account of the objects with which Descriptive Geology is occupied, it must have been felt how difficult it is, in contemplating such facts, to confine ourselves to description and classification. Conjectures and reasonings respecting the causes of the phenomena force themselves upon us at every step; and even influence our classification and nomenclature. Our Descriptive Geology impels us to endeavour to construct a Physical Geology. This close connexion of the two branches of the subject by no means invalidates the necessity of distinguishing them: as in Botany, although the formation of a Natural System necessarily brings us to physiological relations, we still distinguish Systematic from Physiological Botany.

Supposing, however, our Descriptive Geology to be completed, as far as can be done without considering closely the causes by which the strata have been produced, we have now to enter upon the other province of the science, which treats of those causes, and of which we have already spoken, as *Physical Geology.* But before we can treat this department of speculation in a manner suitable to the conditions of science, and to the analogy of other parts of our knowledge, a certain intermediate and preparatory science must be formed, of which we shall now consider the origin and progress.

---

[8] *Brit. Assoc. Report,* 1842, p. 83.

THE

PHILOSOPHY

OF THE

INDUCTIVE SCIENCES,

FOUNDED UPON THEIR HISTORY.

BY WILLIAM WHEWELL, D.D.,

MASTER OF TRINITY COLLEGE, CAMBRIDGE.

A NEW EDITION,

WITH CORRECTIONS AND ADDITIONS, AND AN APPENDIX, CONTAINING

PHILOSOPHICAL ESSAYS PREVIOUSLY PUBLISHED.

IN TWO VOLUMES.

Λαμπάδια ἔχοντες διαδώσουσιν ἀλλήλοις.

LONDON:

JOHN W. PARKER, WEST STRAND.

M.DCCC.XLVII.

Quæ adhuc inventa sunt in Scientiis, ea hujusmodi sunt ut Notionibus Vulgaribus fere subjaceant: ut vero ad interiora et remotiora Naturæ penetretur, necesse est ut tam NOTIONES quam AXIOMATA magis certâ et munitâ viâ a particularibus abstrahantur; atque omnino melior et certior intellectûs adoperatio in usum veniat.

BACON, *Nov. Org.*, Lib. I. Aphor. xviii.

BOOK I.

OF IDEAS IN GENERAL.

Chapter I.

INTRODUCTION.

THE PHILOSOPHY OF SCIENCE, if the phrase were to be understood in the comprehensive sense which most naturally offers itself to our thoughts, would imply nothing less than a complete insight into the essence and conditions of all real knowledge, and an exposition of the best methods for the discovery of new truths. We must narrow and lower this conception, in order to mould it into a form in which we may make it the immediate object of our labours with a good hope of success; yet still it may be a rational and useful undertaking, to endeavour to make some advance towards such a Philosophy, even according to the most ample conception of it which we can form. The present work has been written with a view of contributing, in some measure, however small it may be, towards such an undertaking.

But in this, as in every attempt to advance beyond the position which we at present occupy, our hope of success must depend mainly upon our being able to profit, to the fullest extent, by the progress already made. We may best hope to understand the nature and conditions of real knowledge, by studying the nature and conditions of the most certain and stable portions of knowledge which we already possess : and we are most likely to learn the best methods of discovering truth, by examining how truths, now universally recognized, have really been discovered. Now there do exist among us doctrines of solid and acknowledged certainty, and truths of which the discovery has been received with universal applause. These constitute what we commonly term *Sciences;* and of these bodies of exact and enduring knowledge, we have within our reach so large and varied a collection, that we may examine them, and the history of their formation, with a good prospect of deriving from the study such instruction as we seek. We may best hope to make some progress towards the Philosophy of Science, by employing ourselves upon THE PHILOSOPHY OF THE SCIENCES.

The *Sciences* to which the name is most commonly and unhesitatingly given, are those which are concerned about the material world; whether they deal with the celestial bodies, as the sun and stars, or the earth and its products, or the elements; whether they consider the differences which prevail among such objects, or their origin, or their mutual operation. And in all these Sciences it is familiarly understood and assumed, that their doctrines are obtained by a common process of collecting general truths from particular observed facts, which process is termed *Induction.* It is further assumed that both in these and in other provinces of knowledge, so long as this process is duly and legitimately performed, the results will be real substantial truth. And although this process, with the conditions under which it is legitimate, and the general laws of the formation of Sciences, will hereafter be subjects of discussion in this work, I shall at present so far adopt the assumption of which I speak, as to give to the Sciences from which our lessons are to be collected the name of *Inductive* Sciences. And thus it is that I am led to designate my work as THE PHILOSOPHY OF THE INDUCTIVE SCIENCES.

The views respecting the nature and progress of knowledge, towards which we shall be directed by such a course of inquiry as I have pointed out, though derived from those portions of human knowledge which are more peculiarly and technically termed *Sciences*, will by no means be confined, in their bearing, to the domain of such Sciences as deal with the material world, nor even to the whole range of Sciences now existing. On the contrary, we shall be led to believe that the nature of truth is in all subjects the same, and that its discovery involves, in all cases, the like conditions. On one subject of human speculation after another, man's knowledge assumes that exact and substantial character which leads us to term it *Science;* and in all these cases, whether inert matter or living bodies, whether permanent relations or successive occurrences, be the subject of our attention, we can point out certain universal characters which belong to truth, certain general laws which have regulated its progress among men. And we naturally expect that, even when we extend our range of speculation wider still, when we contemplate the world within us as well as the world without us, when we consider the thoughts and actions of men as well as the motions and operations of unintelligent bodies, we shall still find some general analogies which belong to the essence of truth, and run through the whole intellectual universe. Hence we have reason to trust that a just Philosophy of the Sciences may throw light upon the nature and extent of our knowledge in every department of human speculation. By considering what is the real import of our acquisitions, where they are certain and definite, we may learn something respecting the difference between true knowledge and its precarious or illusory semblances; by examining the steps by which such acquisitions have been made, we may discover the conditions under which truth is to be obtained; by tracing the boundary-line between our knowledge and our ignorance, we may ascertain in some measure the extent of the powers of man's understanding.

But it may be said, in such a design there is nothing new; these are objects at which inquiring men have often before aimed. To determine the difference between real and imaginary knowledge, the conditions under which we arrive at truth, the range of the powers of the human mind, has been a favourite employment of speculative men from the earliest to the most recent times. To inquire into the original, certainty, and compass of man's knowledge, the limits of his capacity, the strength and weakness of his reason, has been the professed purpose of many of the most conspicuous and valued labours of the philosophers of all periods up to our own day. It may appear, therefore, that there is little necessity to add one more to these numerous essays; and little hope that any new attempt will make any very important addition to the stores of thought upon such questions, which have been accumulated by the profoundest and acutest thinkers of all ages.

To this I reply, that without at all disparaging the value or importance of the labours of those who have previously written respecting the foundations and conditions of human knowledge, it may still be possible to add something to what they have done. The writings of all great philosophers, up to our own time, form a series which is not yet terminated. The books and systems of philosophy which have, each in its own time, won the admiration of men, and exercised a powerful influence upon their thoughts, have had each its own part and functions in the intellectual history of the world; and other labours which shall succeed these may also have their proper office and useful effect. We may not be able to do much, and yet still it may be in our power to effect something. Perhaps the very advances made by former inquirers may have made it possible for us, at present, to advance still further. In the discovery of truth, in the developement of man's mental powers and privileges, each generation has its assigned part; and it is for us to endeavour to perform our portion of this perpetual task of our species. Although the terms which describe our undertaking may be the same which have often been employed by previous writers to express their purpose, yet our position is different from theirs, and thus the result may be different too. We have, as they had, to run our appropriate course of speculation with the exertion of our best powers; but our course lies in a more advanced part of the great line along which Philosophy travels from age to age. However familiar and old, therefore, be the design of such a work as this, the execution may have, and if it be performed in a manner suitable to the time, will have, something that is new and not unimportant.

Indeed, it appears to be absolutely necessary, in order to check the prevalence of grave and pernicious errour, that the doctrines which are taught concerning the foundations of human knowledge and the powers of the human mind, should be from time to time revised and corrected or extended. Erroneous and partial views are promulgated and accepted; one portion of the truth is insisted upon to the undue exclusion of another; or principles true in themselves are exaggerated till they produce on men's minds the effect of falsehood. When evils of this kind have grown to a serious height, a *Reform* is requisite. The faults of the existing systems must be remedied by correcting what is wrong, and supplying what is wanting. In such cases, all the merits and excellencies of the labours of the preceding times do not supersede the necessity of putting forth new views suited to the emergency which has arrived. The new form which errour has assumed makes it proper to endeavour to give a new and corresponding form to truth. Thus the mere progress of time, and the natural growth of opinion from one stage to another, leads to the production of new systems and forms of philosophy. It will be found, I think, that some of the doctrines now most widely prevalent respecting the foundations and nature of truth are of such a kind that a Reform is needed. The present age seems, by many indications, to be called upon to seek a sounder Philosophy of Knowledge than is now current among us. To contribute towards such a Philosophy is the object of the present work. The work is, therefore, like all works which take into account the most recent forms of speculative doctrine, invested with a certain degree of novelty in its aspect and import, by the mere time and circumstances of its appearance.

But, moreover, we can point out a very important peculiarity by which this work is, in its design, distinguished from preceding essays on like subjects; and this difference appears to be of such a kind as may well entitle us to expect some substantial addition to our knowledge as the result of our labours. The peculiarity of which I speak has already been announced;—it is this: that we purpose to collect our doctrines concerning the nature of knowledge, and the best mode of acquiring it, from a contemplation of the Structure and History of those Sciences (the Material Sciences), which are universally recognized as the clearest and surest examples of knowledge and of discovery. It is by surveying and studying the whole mass of such Sciences, and the various steps of their progress, that we now hope to approach to the true Philosophy of Science.

Now this, I venture to say, is a new method of pursuing the philosophy of human knowledge. Those who have hitherto endeavoured to explain the nature of knowledge, and the process of discovery, have, it is true, often illustrated their views by adducing special examples of truths which they conceived to be established, and by referring to the mode of their establishment. But these examples have, for the most part, been taken at random, not selected according to any principle or system. Often they have involved doctrines so precarious or so vague that they confused rather than elucidated the subject; and instead of a single difficulty,— What is the nature of Knowledge? these attempts at illustration introduced two,—What was the true analysis of the Doctrines thus adduced? and,—Whether they might safely be taken as types of real Knowledge?

This has usually been the case when there have been adduced, as standard examples of the formation of human knowledge, doctrines belonging to supposed sciences other than the material sciences; doctrines, for example, of Political Economy, or Philology, or Morals, or the Philosophy of the Fine Arts. I am very far from thinking that, in regard to such subjects, there are no important truths hitherto established: but it would seem that those truths which have been obtained in these provinces of knowledge, have not yet been fixed by means of distinct and permanent phraseology, and sanctioned by universal reception, and formed into a connected system, and traced through the steps of their gradual discovery and establishment, so as to make them instructive examples of the nature and progress of truth in general. Hereafter we trust to be able to show that the progress of moral, and political, and philological, and other knowledge, is governed by the same laws as that of physical science. But since, at present, the former class of subjects are full of controversy, doubt, and obscurity, while the latter consist of undisputed truths clearly understood and expressed, it may be considered a wise procedure to make the latter class of doctrines the basis of our speculations. And on the having taken this course, is, in a great measure, my hope founded, of obtaining valuable truths which have escaped preceding inquirers.

But it may be said that many preceding writers on the nature and progress of knowledge have taken their examples abundantly from the Physical Sciences. It would be easy to point out admirable works, which have appeared during the present and former generations, in which instances of discovery, borrowed from the Physical Sciences, are introduced in a manner most happily instructive. And to the works in which this has been done, I gladly give my most cordial admiration. But at the same time I may venture to remark that there still remains a difference between my design and theirs: and that I use the Physical Sciences as exemplifications of the general progress of knowledge in a manner very materially different from the course which is followed in works such as are now referred to. For the conclusions stated in the present work, respecting knowledge and discovery, are drawn from *a connected and systematic survey of the whole range of Physical Science and its History*; whereas, hitherto, philosophers have contented themselves with adducing detached examples of scientific doctrines, drawn from one or two departments of science. So long as we select our examples in this arbitrary and limited manner, we lose the best part of that philosophical instruction, which the sciences are fitted to afford when we consider them as all members of one series, and as governed by rules which are the same for all. Mathematical and chemical truths, physical and physiological doctrines, the sciences of classification and of causation, must alike be taken into our account, in order that we may learn what are the general characters of real knowledge. When our conclusions assume so comprehensive a shape that they apply to a range of subjects so vast and varied as these, we may feel some confidence that they represent the genuine form of universal and permanent truth. But if our exemplification is of a narrower kind, it may easily cramp and disturb our philosophy. We may, for instance, render our views of truth and its evidence so rigid and confined as to be quite worthless, by founding them too much on the contemplation of mathematical truth. We may overlook some of the most important steps in the general course of discovery, by fixing our attention too exclusively upon some one conspicuous group of discoveries, as, for instance, those of Newton. We may misunderstand the nature of physiological discoveries, by attempting to force an analogy between them and discoveries of mechanical laws, and by not attending to the intermediate sciences which fill up the vast interval between these extreme terms in the series of material sciences. In these and in many other ways, a partial and arbitrary reference to the material sciences in our inquiry into human knowledge may mislead us; or at least may fail to give us those wider views, and that deeper insight, which should result from a systematic study of the whole range of sciences with this particular object.

The design of the following work, then, is to form a Philosophy of Science, by analyzing the substance and examining the progress of the existing body of the sciences. As a preliminary to this undertaking, a survey of the history of the sciences was necessary. This, accordingly, I have already performed; and the result of the labour thus undertaken has been laid before the public as a *History of the Inductive Sciences.*

In that work I have endeavoured to trace the steps by which men acquired each main portion of that knowledge on which they now look with so much confidence and satisfaction. The events which that History relates, the speculations and controversies which are there described, and discussions of the same kind, far more extensive, which are there omitted, must all be taken into our account at present, as the prominent and standard examples of the circumstances which attend the progress of knowledge. With so much of real historical fact before us, we may hope to avoid such views of the processes of the human mind as are too partial and limited, or too vague and loose, or too abstract and unsubstantial, to represent fitly the real forms of discovery and of truth.

Of former attempts, made with the same view of tracing the conditions of the progress of knowledge, that of Bacon is perhaps the most conspicuous: and his labours on this subject were opened by his book on the *Advancement of Learning*, which contains, among other matter, a survey of the then existing state of knowledge. But this review was undertaken rather with the object of ascertaining in what quarters future advances were to be hoped for, than of learning by what means they were to be made. His examination of the domain of human knowledge was conducted rather with the view of discovering what remained undone, than of finding out how so much had been done. Bacon's survey was made for the purpose of tracing the boundaries, rather than of detecting the principles of knowledge. "I will now attempt," he says*, "to make a general and faithful perambulation of learning, with an inquiry what parts thereof lie fresh and waste, and not improved and converted by the industry of man; to the end that such a plot made and recorded to memory, may both minister

* *Advancement of Learning*, b. i. p. 74.

light to any public designation, and also serve to excite voluntary endeavours." Nor will it be foreign to our scheme also hereafter to examine with a like purpose the frontier-line of man's intellectual estate. But the object of our perambulation in the first place, is not so much to determine the extent of the field, as the sources of its fertility. We would learn by what plan and rules of culture, conspiring with the native forces of the bounteous soil, those rich harvests have been produced which fill our garners. Bacon's maxims, on the other hand, respecting the mode in which he conceived that knowledge was thenceforth to be cultivated, have little reference to the failures, still less to the successes, which are recorded in his Review of the learning of his time. His precepts are connected with his historical views in a slight and unessential manner. His Philosophy of the Sciences is not collected from the Sciences which are noticed in his survey. Nor, in truth, could this, at the time when he wrote, have easily been otherwise. At that period, scarce any branch of physics existed as a science, except Astronomy. The rules which Bacon gives for the conduct of scientific researches are obtained, as it were, by divination, from the contemplation of subjects with regard to which no sciences as yet were. His instances of steps rightly or wrongly made in this path, are in a great measure cases of his own devising. He could not have exemplified his Aphorisms by references to treatises then extant, on the laws of nature; for the constant burden of his exhortation is, that men up to his time had almost universally followed an erroneous course. And however we may admire the sagacity with which he pointed the way along a better path, we have this great advantage over him;—that we can interrogate the many travellers who since his time have journeyed on this road. At the present day, when we have under our notice so many sciences, of such wide extent, so well established; a Philosophy of the Sciences ought, it must seem, to be founded, not upon conjecture, but upon an examination of many instances;—should not consist of a few vague and unconnected maxims, difficult and doubtful in their application, but should form a system of which every part has been repeatedly confirmed and verified.

This accordingly it is the purpose of the present work to attempt. But I may further observe, that as my hope of making any progress in this undertaking is founded upon the design of keeping constantly in view the whole result of the past history and present condition of science, I have also been led to draw my lessons from my examples in a manner more systematic and regular, as appears to me, than has been done by preceding writers. Bacon, as I have just said, was led to his maxims for the promotion of knowledge by the sagacity of his own mind, with little or no aid from previous examples. Succeeding philosophers may often have gathered useful instruction from the instances of scientific truths and discoveries which they adduced, but their conclusions were drawn from their instances casually and arbitrarily. They took for their moral any which the story might suggest. But such a proceeding as this cannot suffice for us, whose aim is to obtain a consistent body of philosophy from a contemplation of the whole of Science and its History. For our purpose it is necessary to resolve scientific truths into their conditions and ingredients, in order that we may see in what manner each of these has been and is to be provided, in the cases which we may have to consider. This accordingly is necessarily the first part of our task:—*to analyze Scientific Truth into its Elements.* This attempt will occupy the earlier portion of the present work; and will necessarily be somewhat long, and perhaps, in many parts, abstruse and uninviting. The risk of such an inconvenience is inevitable; for the inquiry brings before us many of the most dark and entangled questions in which men have at any time busied themselves. And even if these can now be made clearer and plainer than of yore, still they can be made so only by means of mental discipline and mental effort. Moreover this analysis of scientific truth into its elements contains much, both in its principles and in its results, different from the doctrines most generally prevalent among us in recent times: but on that very account this analysis is an essential part of the doctrines which I have now to lay before the reader: and I must therefore crave his indulgence towards any portion of it which may appear to him obscure or repulsive.

There is another circumstance which may tend to make the present work less pleasing than others on the same subject, in the nature of the examples of human knowledge to which I confine myself; all my instances being, as I have said, taken from the material sciences. For the truths belonging to these sciences are, for the most part, neither so familiar nor so interesting to the bulk of readers as those doctrines which belong to some other subjects. Every general proposition concerning politics or morals at once stirs up an interest in men's bosoms, which makes them listen with curiosity to the attempts to trace it to its origin and foundation. Every rule of art or language brings before the mind of cultivated men subjects of familiar and agreeable thought, and is dwelt upon with pleasure for its own sake, as well as on account of the philosophical lessons which it may convey. But the curiosity which regards the truths of physics or chemistry, or even of physiology and astronomy, is of a more limited and less animated kind.

Hence, in the mode of inquiry which I have prescribed to myself, the examples which I have to adduce will not amuse and relieve the reader's mind as much as they might do, if I could allow myself to collect them from the whole field of human knowledge. They will have in them nothing to engage his fancy, or to warm his heart. I am compelled to detain the listener in the chilly air of the external world, in order that we may have the advantage of full daylight.

But although I cannot avoid this inconvenience, so far as it is one, I hope it will be recollected how great are the advantages which we obtain by this restriction. We are thus enabled to draw all our conclusions from doctrines which are universally allowed to be eminently certain, clear, and definite. The portions of knowledge to which I refer are well known, and well established among men. Their names are familiar, their assertions uncontested. Astronomy and Geology, Mechanics and Chemistry, Optics and Acoustics, Botany and Physiology, are each recognized as large and substantial collections of undoubted truths. Men are wont to dwell with pride and triumph on the acquisitions of knowledge which have been made in each of these provinces; and to speak with confidence of the certainty of their results. And all can easily learn in what repositories these treasures of human knowledge are to be found. When, therefore, we begin our inquiry from such examples, we proceed upon a solid foundation. With such a clear ground of confidence, we shall not be met with general assertions of the vagueness and uncertainty of human knowledge; with the question, What truth is, and How we are to recognize it; with complaints concerning the hopelessness and unprofitableness of such researches. We have, at least, a definite problem before us. We have to examine the structure and scheme, not of a shapeless mass of incoherent materials, of which we doubt whether it be a ruin or a natural wilderness, but of a fair and lofty palace, still erect and tenanted, where hundreds of different apartments belong to a common plan, where every generation adds something to the extent and magnificence of the pile. The certainty and the constant progress of science are things so unquestioned, that we are at least engaged in an intelligible inquiry, when we are examining the grounds and nature of that certainty, the causes and laws of that progress.

To this enquiry, then, we now proceed. And in entering upon this task, however our plan or our principles may differ from those of the eminent philosophers who have endeavoured, in our own or in former times, to illustrate or enforce the philosophy of science, we most willingly acknowledge them as in many things our leaders and teachers. Each reform must involve its own peculiar principles, and the result of our attempts, so far as they lead to a result, must be, in some respects, different from those of former works. But we may still share with the great writers who have treated this subject before us, their spirit of hope and trust, their reverence for the dignity of the subject, their belief in the vast powers and boundless destiny of man. And we may once more venture to use the words of hopeful exhortation, with which the greatest of those who have trodden this path encouraged himself and his followers when he set out upon his way.

" Concerning ourselves we speak not; but as touching the matter which we have in hand, this we ask;—that men deem it not to be the setting up an Opinion, but the performing of a Work: and that they receive this as a certainty; that we are not laying the foundations of any sect or doctrine, but of the profit and dignity of mankind. Furthermore, that being well disposed to what shall advantage themselves, and putting off factions and prejudices, they take common counsel with us, to the end that being by these our aids and appliances freed and defended from wanderings and impediments, they may lend their hands also to the labours which remain to be performed: and yet further, that they be of good hope; neither imagine to themselves this our Reform as something of infinite dimension, and beyond the grasp of mortal man, when in truth it is the end and true limit of infinite errour; and is by no means unmindful of the condition of mortality and humanity, not confiding that such a thing can be carried to its perfect close in the space of one single age, but assigning it as a task to a succession of generations."

CHAPTER II.

OF THE FUNDAMENTAL ANTITHESIS OF PHILOSOPHY.

SECT. 1.—*Thoughts and Things.*

In order that we may do something towards determining the nature and conditions of human knowledge, (which I have already stated as the purpose of this work,) I shall have to refer to an antithesis or opposition, which is familiar and generally recognized, and in which the distinction of the things opposed to each other is commonly considered very clear and plain. I shall have to attempt to make this opposition sharper and stronger than it is usually conceived, and yet to shew that the distinction is far from being so clear and definite as it is usually assumed to be: I shall have to point the contrast, yet shew that the things which are contrasted cannot be separated:—I must explain that the anti-thesis is constant and essential, but yet that there is no fixed and permanent line dividing its members. I may thus appear, in different parts of my discussion, to be proceeding in opposite directions, but I hope that the reader who gives me a patient attention will see that both steps lead to the point of view to which I wish to lead him.

The antithesis or opposition of which I speak is denoted, with various modifications, by various pairs of terms: I shall endeavour to show the connexion of these different modes of expression, and I will begin with that form which is the simplest and most idiomatic.

The simplest and most idiomatic expression of the antithesis to which I refer is that in which we oppose to each other THINGS and THOUGHTS. The opposition is familiar and plain. Our Thoughts are something which belongs to ourselves; something which takes place within us; they are what *we* think; they are actions of our minds. Things, on the contrary, are something different from ourselves and independent of us; something which is without us; they *are;* we see them, touch them, and thus know that they exist; but we do not make them by seeing or touching them, as we make our *Thoughts* by thinking them; we are passive, and *Things* act upon our organs of perception.

Now what I wish especially to remark is this: that in all human KNOWLEDGE both Thoughts and Things are concerned. In every part of my knowledge there must be some *thing* about which I know, and an internal act of *me* who know. Thus, to take simple yet definite parts of our knowledge, if I know that a solar year consists of 365 days, or a lunar month of 30 days, I know some-thing about the sun or the moon; namely, that those objects perform certain revolutions and go through certain changes, in those numbers of days; but I count such numbers and conceive such revolutions and changes by acts of my own thoughts. And both these elements of my knowledge are indispensable. If there were not such external Things as the sun and the moon I could not have any knowledge of the progress of time as marked by them. And however regular were the motions of the sun and moon, if I could not count their appearances and combine their changes into a cycle, or if I could not understand this when done by other men, I could not know anything about a year or a month. In the former case I might be conceived as a human being, possessing the human powers of thinking and reckoning, but kept in a dark world with nothing to mark the progress of existence. The latter is the case of brute animals, which see the sun and moon, but do not know how many days make a month or a year, because they have not human powers of thinking and reckoning.

The two elements which are essential to our knowledge in the above cases, are necessary to human knowledge in all cases. In all cases, Knowledge implies a combination of Thoughts and Things. Without this combination, it would not be Knowledge. Without Thoughts, there could be no connexion; without Things, there could be no reality. Thoughts and Things are so intimately combined in our Knowledge, that we do not look upon them as distinct. One single act of the mind involves them both; and their contrast disappears in their union.

But though Knowledge requires the union of these two elements, Philosophy requires the separation of them, in order that the nature and structure of Knowledge may be seen. Therefore I begin by considering this separation. And I now proceed to speak of another way of looking at the antithesis of which I have spoken;

and which I may, for the reasons which I have just mentioned, call the FUNDAMENTAL ANTITHESIS OF PHILOSOPHY.

SECT. 2.—*Necessary and Experiential Truths.*

MOST persons are familiar with the distinction of *necessary* and *contingent* truths. The former kind are Truths which cannot but be true; as that 19 and 11 make 30 ;—that parallelograms upon the same base and between the same parallels are equal:—that all the angles in the same segment of a circle are equal. The latter are Truths which *it happens* (*contingit*) are true; but which, for any thing which we can see, might have been otherwise; as that a lunar month contains 30 days, or that the stars revolve in circles round the pole. The latter kind of Truths are learnt by experience, and hence we may call them *Truths of Experience*, or, for the sake of convenience, *Experiential* Truths, in contrast with Necessary Truths.

Geometrical propositions are the most manifest examples of Necessary Truths. All persons who have read and understood the elements of geometry, know that the propositions above stated (that parallelograms upon the same base and between the same parallels are equal; that all the angles in the same segment of a circle are equal,) are necessarily true; not only they *are* true, but they *must be* true. The meaning of the terms being understood, and the proof being gone through, the truth of the propositions must be assented to. We learn these propositions to be true by demonstrations deduced from definitions and axioms ; and when we have thus learnt them, we see that they could not be otherwise. In the same manner, the truths which concern numbers are necessary truths: 19 and 11 not only *do* make 30, but *must* make that number, and cannot make anything else.

In the same manner, it is a necessary truth that half the sum of two numbers added to half their difference is equal to the greater number.

It is easy to find examples of Experiential Truths;—propositions which we know to be true, but know by experience only. We know, in this way, that salt will dissolve in water; that plants cannot live without light; —in short, we know in this way all that we do know in chemistry, physiology, and the material sciences in general. I take the *Sciences* as my examples of human knowledge, rather than the common truths of daily life, or moral or political truths; because, though the latter are more generally interesting, the former are much more definite and certain, and therefore better starting-points for our speculations, as I have already said. And we may take elementary astronomical truths as the most familiar examples of Experiential Truths in the domain of science.

With these examples, the distinction of Necessary and Experiential Truths is, I hope, clear. The former kind, we see to be true by thinking about them, and see that they could not be otherwise. The latter kind, men could never have discovered to be true without looking at them; and having so discovered them, still no one will pretend to say they might not have been otherwise. For aught we can see, the astronomical truths which express the motions and periods of the sun, moon and stars, might have been otherwise. If we had been placed in another part of the solar system, our experiential truths respecting days, years, and the motions of the heavenly bodies, would have been other than they are, as we know from astronomy itself.

It is evident that this distinction of Necessary and Experiential Truths involves the same antithesis which we have already considered;—the antithesis of Thoughts and Things. Necessary Truths are derived from our own Thoughts: Experiential Truths are derived from our observation of Things about us. The opposition of Necessary and Experiential Truths is another aspect of the Fundamental Antithesis of Philosophy.

SECT. 3.—*Deduction and Induction.*

I HAVE already stated that geometrical truths are established by demonstrations *deduced* from definitions and axioms. The term *Deduction* is specially applied to such a course of demonstration of truths from definitions and axioms. In the case of the parallelograms upon the same base and between the same parallels, we prove certain triangles to be equal, by supposing them placed so that their two bases have the same extremities; and hence, referring to an Axiom respecting straight lines, we infer that the bases coincide. We combine these equal triangles with other equal spaces, and in this way make up both the one and the other of the parallelograms, in such a manner as to shew that they are equal. In this manner, going on step by step, deducing the equality of the triangles from the axiom, and the equality of the parallelograms from that of the triangles, we travel to the conclusion. And this process of successive deduction is the scheme of all geometrical proof. We begin with Definitions of the notions which we reason about, and with Axioms, or self-evident truths, respecting these notions; and we get, by reasoning from these, other truths which are demonstratively evident; and from these truths again, others of the same kind, and so on. We begin with our own Thoughts, which supply us with Axioms to start from; and we reason from these, till we come to propositions which are applicable to the Things about us; as for instance, the propositions respecting circles and spheres are applicable to the motions of the heavenly bodies.   This is *Deduction,* or *Deductive Reasoning.*

Experiential truths are acquired in a very different way.   In order to obtain such truths, we begin with Things.   In order to learn how many days there are in a year, or in a lunar month, we must begin by observing the sun and the moon.   We must observe their changes day by day, and try to make the cycle of change fit into some notion of number which we supply from our own Thoughts.   We shall find that a cycle of 30 days nearly will fit the changes of phase of the moon;—that a cycle of 365 days nearly will fit the changes of daily motion of the sun.   Or, to go on to experiential truths of which the discovery comes within the limits of the history of science—we shall find (as Hipparchus found) that the unequal motion of the sun among the stars, such as observation shews it to be, may be fitly represented by the notion of an *eccentric;*—a circle in which the sun has an equable annual motion, the spectator not being in the center of the circle.   Again, in the same manner, at a later period, Kepler started from more exact observations of the sun, and compared them with a supposed motion in a certain ellipse; and was able to shew that, not a circle about an eccentric point, but an ellipse, supplied the mode of conception which truly agreed with the motion of the sun about the earth; or rather, as Copernicus had already shewn, of the earth about the sun.   In such cases, in which truths are obtained by beginning from observation of external things and by finding some notion with which the Things, as observed, agree, the truths are said to be obtained by *Induction.*   The process is an *Inductive Process.*

The contrast of the Deductive and Inductive process is obvious.   In the former, we proceed at each step from general truths to particular applications of them;

in the latter, from particular observations to a general truth which includes them. In the former case we may be said to reason *downwards,* in the latter case, *upwards;* for general notions are conceived as standing above particulars. Necessary truths are proved, like arithmetical sums, by adding together the portions of which they consist. An inductive truth is proved, like the guess which answers a riddle, by its agreeing with the facts described. Demonstation is irresistible in its effect on the belief, but does not produce surprize, because all the steps to the conclusion are exhibited, before we arrive at the conclusion. Inductive inference is not demonstrative, but it is often more striking than demonstrative reasoning, because the intermediate links between the particulars and the inference are not shown. Deductive truths are the results of relations among our own Thoughts. Inductive Truths are relations which we discern among existing Things; and thus, this opposition of Deduction and Induction is again an aspect of the Fundamental Antithesis already spoken of.

SECT. 4.—*Theories and Facts.*

GENERAL experiential Truths, such as we have just spoken of, are called *Theories,* and the particular observations from which they are collected, and which they include and explain, are called *Facts.* Thus Hipparchus's doctrine, that the sun moves in an eccentric about the earth, is *his Theory* of the Sun, or the *Eccentric Theory.* The doctrine of Kepler, that the Earth moves in an Ellipse about the Sun, is *Kepler's Theory* of the Earth, the Elliptical Theory. Newton's doctrine that this elliptical motion of the Earth about the Sun is produced and governed by the Sun's attraction upon the Earth, is the *Newtonian* theory, the *Theory of Attraction.* Each of these Theories was accepted, because it included, connected and explained the *Facts;* the Facts being, in the two former cases, the motions of the Sun as observed; and in the other case, the elliptical motion of the Earth as known by Kepler's Theory. This antithesis of *Theory* and *Fact* is included in what has just been said of Inductive Propositions. A Theory is an Inductive Proposition, and the Facts are the particular observations from which, as I have said, such Propositions are inferred by Induction. The Antithesis of Theory and Fact implies the fundamental Antithesis of Thoughts and Things; for a Theory (that is, a true Theory) may be described as a Thought which is contemplated distinct from Things and seen to agree with them; while a Fact is a combination of our Thoughts with Things in so complete agreement that we do not regard them as separate.

Thus the antithesis of Theory and Fact involves the antithesis of Thoughts and Things, but is not identical with it. Facts involve Thoughts, for we know Facts only by thinking about them. The Fact that the year consists of 365 days; the Fact that the month consists of 30 days, cannot be known to us, except we have the Thoughts of Time, Number and Recurrence. But these Thoughts are so familiar, that we have the Fact in our mind as a simple Thing without attending to the Thought which it involves. When we mould our Thoughts into a Theory, we consider the Thought as distinct from the Facts; but yet, though distinct, not independent of them; for it is a true Theory, only by including and agreeing with the Facts.

SECT. 5.—*Ideas and Sensations.*

WE have just seen that the antithesis of Theory and Fact, although it involves the antithesis of Thoughts and Things, is not identical with it. There are other modes of expression also, which involve the same Fundamental Antithesis, more or less modified. Of these, the pair of words which in their relations appear to separate the members of the antithesis most distinctly are *Ideas* and *Sensations*. We see and hear and touch external things, and thus perceive them by our senses; but in perceiving them, we connect the impressions of sense according to relations of space, time, number, likeness, cause, &c. Now some at least of these kinds of connexion, as space, time, number, may be contemplated distinct from the things to which they are applied; and so contemplated, I term them *Ideas*. And the other element, the impressions upon our senses which they connect, are called *Sensations*.

I term space, time, cause, &c., *Ideas*, because they are general relations among our sensations, apprehended by an act of the mind, not by the senses simply. These relations involve something beyond what the senses alone could furnish. By the sense of sight we see various shades and colours and shapes before us, but the *outlines* by which they are separated into distinct objects of definite forms, are the work of the mind itself. And again, when we conceive visible things, not only as surfaces of a certain form, but as *solid bodies*, placed at various distances in space, we again exert an act of the mind upon them. When we see a body move, we see it move in a path or *orbit*, but this orbit is not itself seen; it is constructed by the mind. In like manner when we see the motions of a needle towards a magnet, we do not *see* the attraction or force which produces the effects; but we infer the force, by having in our minds the Idea of Cause. Such acts of thought, such *Ideas*, enter into our perceptions of external things.

But though our perceptions of external things involve some act of the mind, they must involve something else besides an act of the mind. If we must exercise an act of thought in order to see force exerted, or orbits described by bodies in motion, or even in order to see bodies existing in space, and to distinguish one kind of object from another, still the act of thought alone does not make the bodies. There must be something besides, *on which* the thought is exerted. A colour, a form, a sound, are not produced by the mind, however they may be moulded, combined, and interpreted by our mental acts. A philosophical poet has spoken of

> All the world
> Of eye and ear, both what they half create,
> And what perceive.

But it is clear, that though they *half* create, they do not wholly create: there must be an external world of colour and sound to give impressions to the eye and ear, as well as internal powers by which we perceive what is offered to our organs. The mind is in some way passive as well as active: there are objects without as well as faculties within;—Sensations, as well as acts of Thought.

Indeed this is so far generally acknowledged, that according to common apprehension, the mind is passive *rather* than active in acquiring the knowledge which it receives concerning the material world. Its sensations are generally considered more distinct than its operations. The world without is held to be more clearly real than the faculties within. That there is something different from ourselves, something external to us, something independent of us, something which no act of our minds can make or can destroy, is held by all men to be at least as evident, as that our minds can exert any effectual process in modifying and appreciating the impressions made upon them. Most persons are more likely to doubt whether the mind be always actively applying Ideas to the objects which it perceives, than whether it perceive them passively by means of Sensations.

But yet a little consideration will show us that an activity of the mind, and an activity according to certain Ideas, is requisite in all our knowledge of external objects. We see objects, of various solid forms, and at various distances from us. But we do not thus perceive them by sensation alone. Our visual impressions cannot, of themselves, convey to us a knowledge of solid form, or of distance from us. Such knowledge is inferred from what we see:—inferred by conceiving the objects as existing in space, and by applying to them the Idea of Space. Again:—day after day passes, till they make up a year: but we do not know that the days are 365, except we count them; and thus apply to them our Idea of Number. Again:—we see a needle drawn to a magnet: but, in truth, the *drawing* is what we cannot see. We see the needle move, and infer the attraction, by applying to the fact our Idea of Force, as the cause of motion. Again:— we see two trees of different kinds; but we cannot know that they are so, except by applying to them our Idea of the resemblance and difference which makes kinds. And thus Ideas, as well as Sensations, necessarily enter into all our knowledge of objects: and these two words express, perhaps more exactly than any of the pairs before mentioned, that Fundamental Antithesis, in the union of which, as I have said, all knowledge consists.

SECT 6.—*Reflexion and Sensation.*

IT will hereafter be my business to show what the Ideas are, which thus enter into our knowledge; and how each Idea has been, as a matter of historical fact, introduced into the Science to which it especially belongs. But before I proceed to do this, I will notice some other terms, besides the phrases already noticed, which have a reference, more or less direct, to the Fundamental Antithesis of Ideas and Sensations. I will mention some of these, in order that if they should come under the reader's notice, he may not be perplexed as to their bearing upon the view here presented to him.

The celebrated doctrine of Locke, that all our "Ideas," (that is, in his use of the word, all our objects of thinking,) come from Sensation or Reflexion, will naturally occur to the reader as connected with the antithesis of which I have been speaking. But there is a great difference between Locke's account of Sensation and Reflexion, and our view of Sensation and Ideas. He is speaking of the origin of our knowledge;—we, of its nature and composition. He is content to say that all the knowledge which we do not receive directly by Sensation, we obtain by Reflex Acts of the mind, which make up his Reflexion. But we hold that there is no Sensation without an act of the mind, and that the mind's activity is not only reflexly exerted upon itself, but directly upon objects, so as to perceive in them connexions and relations which are not Sensations. He is content to put together, under the name of Reflexion, everything in our knowledge which is not Sensation: we are to attempt to analyze all that is not Sensation; not only to say it consists of Ideas, but to point out what those Ideas are, and to show the mode in which each of them enters into our knowledge. His purpose was, to prove that there are no Ideas, except the reflex acts of the mind: our endeavour will be to show that the acts of the mind, both direct and reflex, are governed by certain Laws, which may be conveniently termed Ideas. His procedure was, to deny that any knowledge could be derived from the mind alone: our course will be, to show that in every part of our most certain and exact knowledge, those who have added to our knowledge in every age have referred to principles which the mind itself supplies. I do not say that my view is contrary to his: but it is altogether different from his. If I grant that all our knowledge comes from Sensation and Reflexion, still my task then is only begun; for I want further to determine, in each science, what portion comes, not from mere Sensation, but from those Ideas by the aid of which either Sensation or Reflexion can lead to Science.

Locke's use of the word "idea" is, as the reader will perceive, different from ours. He uses the word, as he says, which "serves best to stand for whatsoever is the object of the understanding when a man thinks." "I have used it," he adds, "to express whatever is meant by *phantasm, notion, species*, or whatever it is to which the mind can be employed about in thinking." It might be shown that this separation of the *mind itself* from the ideal *objects* about which it is employed in thinking, may lead to very erroneous results. But it may suffice to observe that we use the word *Ideas*, in the manner already explained, to express that element, supplied by the mind itself, which must be combined with Sensation in order to produce knowledge. For us, Ideas are not Objects of Thought, but rather Laws of Thought. Ideas are not synonymous with Notions; they are Principles which give to our Notions whatever they contain of truth. But our use of the term *Idea* will be more fully explained hereafter.

Sect. 7—*Subjective and Objective.*

The Fundamental Antithesis of Philosophy of which I have to speak has been brought into great prominence in the writings of modern German philosophers, and has conspicuously formed the basis of their systems. They have indicated this antithesis by the terms *subjective* and *objective*. According to the technical language of old writers, a thing and its qualities are described as *subject* and *attributes;* and thus a man's faculties and acts are attributes of which he is the *subject.* The mind is the *subject* in which ideas inhere. Moreover, the man's faculties and acts are employed upon external *objects;* and from objects all his sensations arise. Hence the part of a man's knowledge which belongs to his own mind, is *subjective:* that which flows in upon him from the world external to him, is *objective.* And as in man's contemplation of nature, there is always some act of thought which depends upon himself, and some matter of thought which is independent of him, there is, in every part of his knowledge, a subjective and an objective element. The combination of the two elements, the subjective or ideal, and the objective or observed, is necessary, in order to give us any insight into the laws of nature. But different persons, according to their mental habits and constitution, may be inclined to dwell by preference upon the one or the other of these two elements. It may perhaps interest the reader to see this difference of intellectual character illustrated in two eminent men of genius of modern times, Göthe and Schiller.

Göthe himself gives us the account to which I refer, in his history of the progress of his speculations concerning the Metamorphosis of Plants; a mode of viewing their structure by which he explained, in a very striking and beautiful manner, the relations of the different parts of a plant to each other; as has been narrated in the *History of the Inductive Sciences.* Göthe felt a delight in the passive contemplation of nature, unmingled with the desire of reasoning and theorizing; a delight such as naturally belongs to those poets who merely embody the images which a fertile genius suggests, and do not mix with these pictures, judgments and reflexions of their own. Schiller, on the other hand, both by his own strong feeling of the value of a moral purpose in poetry, and by his adoption of a system of metaphysics in which the subjective element was made very prominent, was well disposed to recognize fully the authority of ideas over external impressions.

Göthe for a time felt a degree of estrangement towards Schiller, arising from this contrariety in their views and characters. But on one occasion they fell into discussion on the study of natural history; and Göthe endeavoured to impress upon his companion his persuasion that nature was to be considered, not as composed of detached and incoherent parts, but as active and alive, and unfolding herself in each portion, in virtue of principles which pervade the whole. Schiller objected that no such view of the objects of natural history had been pointed out by observation, the only guide which the natural historians recommended; and was disposed on this account to think the whole of their study narrow and shallow. "Upon this," says Göthe, "I expounded to him, in as lively a way as I could, the metamorphosis of plants, drawing on paper for him, as I proceeded, a diagram to represent that general form of a plant which shows itself in so many and so various transformations. Schiller attended and understood; and, accepting the explanation, he said, 'This is not observation, but an idea.' I replied," adds Göthe, "with some degree of irritation; for the point which separated us was most luminously marked by this expression: but I smothered my vexation, and merely said, 'I was happy to find that I had got ideas without knowing it; nay, that I saw them before my eyes.'" Göthe then goes on to say, that he had been grieved to the very soul by maxims promulgated by Schiller, that no observed fact ever could correspond with an idea. Since he himself loved best to wander in the domain of external observation, he had been led to look with repugnance and hostility upon anything which professed to depend upon ideas. "Yet," he observes, "it occurred to me that if my Observation was identical with his Idea, there must be some common ground on which we might meet." They went on with their mutual explanations, and became intimate and lasting friends. "And thus," adds the poet, "by means of that mighty and interminable controversy between *object* and *subject*, we two concluded an alliance which remained unbroken, and produced much benefit to ourselves and others."

The general diagram of a plant, of which Göthe here speaks, must have been a combination of lines and marks expressing the relations of position and equivalence among the elements of vegetable forms, by which so many of their resemblances and differences may be explained. Such a symbol is not an Idea in that general sense in which we propose to use the term, but is a particular modification of the general Ideas of symmetry, developement, and the like; and we shall hereafter see, according to the phraseology which we shall explain in the next chapter, how such a diagram might express the *ideal conception* of a plant.

The antithesis of *subjective* and *objective* is very familiar in the philosophical literature of Germany and France; nor is it uncommon in any age of our own literature. But though efforts have recently been made to give currency among us to this phraseology, it has not been cordially received, and has been much complained of as not of obvious meaning. Nor is the complaint without ground: for when we regard the mind as the *subject* in which ideas inhere, it becomes for us an *object*, and the antithesis vanishes. We are not so much accustomed to use *subject* in this sense, as to make it a proper contrast to *object*. The combination "*ideal* and *objective*," would more readily convey to a modern reader the opposition which is intended between the ideas of the mind itself, and the objects which it contemplates around it.

To the antitheses already noticed—Thoughts and Things; Necessary and Experiential Truths; Deduction and Induction; Theory and Fact; Ideas and Sensations; Reflexion and Sensation; Subjective and Objective; we may add others, by which distinctions depending more or less upon the fundamental antithesis have been de-noted. Thus we speak of the *internal* and *external* sources of our knowledge; of the world *within* and the world *without* us; of *Man* and *Nature*. Some of the more recent metaphysical writers of Germany have divided the universe into the *Me* and the *Not-me* (Ich and Nicht-ich). Upon such phraseology we may observe, that to have the fundamental antithesis of which we speak really understood, is of the highest consequence to philosophy, but that little appears to be gained by expressing it in any novel manner. The most weighty part of the philosopher's task is to analyze the operations of the mind; and in this task, it can aid us but little to call it, instead of the *mind*, the *subject*, or the *me*.

BOOK II

THE PHILOSOPHY OF THE PURE SCIENCES.

CHAPTER I.

OF THE PURE SCIENCES.

1. ALL external objects and events which we can contemplate are viewed as having relations of Space, Time, and Number; and are subject to the general conditions which these Ideas impose, as well as to the particular laws which belong to each class of objects and occurrences. The special laws of nature, considered under the various aspects which constitute the different sciences, are obtained by a mixed reference to experience and to the fundamental ideas of each science. But besides the sciences thus formed by the aid of special experience, the conditions which flow from those more comprehensive ideas first mentioned, Space, Time, and Number, constitute a body of science, applicable to objects and changes of all kinds, and deduced without recurrence being had to any observation in particular. These sciences, thus unfolded out of ideas alone, unmixed with any reference to the phenomena of matter, are hence termed *Pure* Sciences. The principal sciences of this class are Geometry, Theoretical Arithmetic, and Algebra considered in its most general sense, as the investigation of the relations of space and number by means of general symbols.

2. These Pure Sciences were not included in our survey of the history of the sciences, because they are not *inductive* sciences. Their progress has not consisted in collecting laws from phenomena, true theories from observed facts, and more general from more limited laws; but in tracing the consequences of the ideas themselves, and in detecting the most general and intimate analogies and connexions which prevail among such conceptions as are derivable from the ideas. These sciences have no principles besides definitions and axioms, and no process of proof but *deduction;* this process, however, assuming here a most remarkable character; and exhibiting a combination of simplicity and complexity, of rigour and generality, quite unparalleled in other subjects.

3. The universality of the truths, and the rigour of the demonstrations of these pure sciences, attracted attention in the earliest times; and it was perceived that they offered an exercise and a discipline of the intellectual faculties, in a form peculiarly free from admixture of extraneous elements. They were strenuously cultivated by the Greeks, both with a view to such a discipline, and from the love of speculative truth which prevailed among that people: and the name *mathematics,* by which they are designated, indicates this their character of *disciplinal* studies.

4. As has already been said, the ideas which these sciences involve extend to all the objects and changes which we observe in the external world; and hence the consideration of mathematical relations forms a large portion of many of the sciences which treat of the phenomena and laws of external nature, as Astronomy, Optics, and Mechanics. Such sciences are hence often termed *Mixed Mathematics,* the relations of space and number being, in these branches of knowledge, combined with principles collected from special observation;

while Geometry, Algebra, and the like subjects, which involve no result of experience, are called *Pure Mathematics*.

5. Space, time, and number, may be conceived as *forms* by which the knowledge derived from our sensations is moulded, and which are independent of the differences in the *matter* of our knowledge, arising from the sensations themselves. Hence the sciences which have these ideas for their subject may be termed *Formal Sciences*. In this point of view, they are distinguished from sciences in which, besides these mere formal laws by which appearances are corrected, we endeavour to apply to the phenomena the idea of cause, or some of the other ideas which penetrate further into the principles of nature. We have thus, in the History, distinguished Formal Astronomy and Formal Optics from Physical Astronomy and Physical Optics.

We now proceed to our examination of the Ideas which constitute the foundation of these formal or pure mathematical sciences, beginning with the Idea of Space.

Chapter VII.

OF THE IDEA OF TIME.

1. Respecting the Idea of Time, we may make several of the same remarks which we made concerning

---

* The expression in the first edition was " large objects and extensive spaces." In the text as now given, I state a definite size and extent, within which the sight by itself can judge of position and figure.

The doctrine that we require the assistance of the muscular sense to enable us to perceive space of three dimensions, is not at all inconsistent with this other doctrine, that within the space which is seen by the fixed eye, we perceive the relative positions of points directly by vision, and that, consequently, we have a perception of *visible figure*.

Sir Charles Bell has said, (*Phil. Trans.* 1823, p. 181,) " It appears to me that the utmost ingenuity will be at a loss to devise an explanation of that power by which the eye becomes acquainted with the position and relation of objects, if the sense of muscular activity be excluded which accompanies the motion of the eyeball." But surely we should have no difficulty in perceiving the relation of the sides and angles of a small triangle, placed before the eye, even if the muscles of the eyeball were severed. This subject is resumed B. iv. c. ii. sect. 11.

the idea of space, in order to shew that it is not borrowed from experience, but is a bond of connexion among the impressions of sense, derived from a peculiar activity of the mind, and forming a foundation both of our experience and of our speculative knowledge.

Time is not a notion obtained by experience. Experience, that is, the impressions of sense and our consciousness of our thoughts, gives us various perceptions; and different successive perceptions considered together exemplify the notion of change. But this very connexion of different perceptions,—this successiveness, —presupposes that the perceptions exist *in time*. That things happen either together, or one after the other, is intelligible only by assuming time as the condition under which they are presented to us.

Thus time is a necessary condition in the presentation of all occurrences to our minds. We cannot conceive this condition to be taken away. We can conceive time to go on while nothing happens in it; but we cannot conceive anything to happen while time does not go on.

It is clear from this that time is not an impression derived from experience, in the same manner in which we derive from experience our information concerning the objects which exist, and the occurrences which take place in time. The objects of experience can easily be conceived to be, or not to be:—to be absent as well as present. Time always is, and always is present, and even in our thoughts we cannot form the contrary supposition.

2. Thus time is something distinct from the *matter* or substance of our experience, and may be considered as a necessary *form* which that matter (the experience of change) must assume, in order to be an object of contemplation to the mind. Time is one of the necessary

conditions under which we apprehend the information which our senses and consciousness give us. By considering time as a form which belongs to our power of apprehending occurrences and changes, and under which alone all such experience can be accepted by the mind, we explain the necessity, which we find to exist, of conceiving all such changes as happening in time; and we thus see that time is not a property perceived as existing in objects, or as conveyed to us by our senses; but a condition impressed upon our knowledge by the constitution of the mind itself; involving an act of thought as well as an impression of sense.

3. We showed that space is an idea of the mind, or form of our perceiving power, independent of experience, by pointing out that we possess necessary and universal truths concerning the relations of space, which could never be given by means of experience; but of which the necessity is readily conceivable, if we suppose them to have for their basis the constitution of the mind. There exist also respecting number, many truths absolutely necessary, entirely independent of experience and anterior to it; and so far as the conception of number depends upon the idea of time, the same argument might be used to show that the idea of time is not derived from experience, but is a result of the native activity of the mind: but we shall defer all views of this kind till we come to the consideration of Number.

4. Some persons have supposed that we obtain the notion of time from the perception of motion. But it is clear that the perception of motion, that is, change of place, presupposes the conception of time, and is not capable of being presented to the mind in any other way. If we contemplate the same body as being in different places at different times, and connect these observations, we have the conception of motion, which thus presup-

poses the necessary conditions that existence in time implies. And thus we see that it is possible there should be necessary truths concerning all motion, and consequently, concerning those motions which are the objects of experience; but that the source of this necessity is the Ideas of time and space, which, being universal conditions of knowledge residing in the mind, afford a foundation for necessary truths.

CHAPTER VIII.

OF SOME PECULIARITIES OF THE IDEA OF TIME.

1. THE Idea of Time, like the Idea of Space, offers to our notice some characters which do not belong to our fundamental ideas generally, but which are deserving of remark. These characters are, in some respects, closely similar with regard to time and to space, while, in other respects, the peculiarities of these two ideas are widely different. We shall point out some of these characters.

Time is not a general *abstract* notion collected from experience; as, for example, a certain general conception of the relations of things. For we do not consider particular *times* as examples of Time in general, (as we consider particular causes to be examples of Cause,) but we conceive all particular times to be parts of a single and endless Time. This continually-flowing and endless time is what offers itself to us when we contemplate any series of occurrences. All actual and possible times exist as Parts, in this original and general Time. And since all particular times are considered as derivable from time in general, it is manifest that the notion of time in general cannot be derived from the notions of particular times. The notion of time in general is therefore not a general conception gathered from experience.

2. Time is infinite. Since all actual and possible times exist in the general course of time, this general time must be infinite. All limitation merely divides, and does not terminate, the extent of absolute time. Time has no beginning and no end; but the beginning and the end of every other existence takes place in it.

3. Time, like space, is not only a form of perception, but of *intuition*. We contemplate events as taking place *in* time. We consider its parts as added to one another, and events as filling a larger or smaller extent of such parts. The time which any event takes up is the sum of all such parts, and the relation of the same to time is fully understood when we can clearly see what portions of time it occupies, and what it does not. Thus the relation of known occurrences to time is perceived by intuition; and time is a form of intuition of the external world.

4. Time is conceived as a quantity of one dimension; it has great analogy with a line, but none at all with a surface or solid. Time may be considered as consisting of a series of instants, which are before and after one another; and they have no other relation than this, of *before* and *after*. Just the same would be the case with a series of points taken along a line; each would be after those on one side of it, and before those on another. Indeed the analogy between time, and space of one dimension, is so close, that the same terms are applied to both ideas, and we hardly know to which they originally belong. Times and lines are alike called *long* and *short;* we speak of the *beginning* and *end* of a line; of a *point* of time, and of the *limits* of a portion of duration.

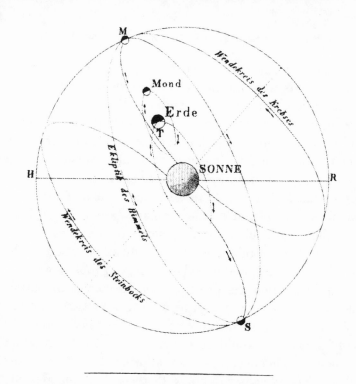

Chapter VIII.

OF THE PARADOX OF UNIVERSAL PROPOSITIONS OBTAINED FROM EXPERIENCE.

1. It was formerly stated* that experience cannot establish any universal or necessary truths. The number of trials which we can make of any proposition is necessarily limited, and observation alone cannot give us any ground of extending the inference to untried cases. Observed facts have no visible bond of necessary connexion, and no exercise of our senses can enable us to discover such connexion. We can never acquire from a mere observation of facts, the right to assert that a proposition is true in all cases, and that it could not be otherwise than we find it to be.

* B. i., c. v. Of Experience.

Yet, as we have just seen in the history of the laws of motion, we may go on collecting our knowledge from observation, and enlarging and simplifying it, till it approaches or attains to complete universality and seeming necessity. Whether the laws of motion, as we now know them, can be rigorously traced to an absolute necessity in the nature of things, we have not ventured absolutely to pronounce. But we have seen that some of the most acute and profound mathematicians have believed that, for these laws of motion, or some of them, there was such a demonstrable necessity compelling them to be such as they are, and no other. Most of those who have carefully studied the principles of Mechanics will allow that some at least of the primary laws of motion approach very near to this character of necessary truth; and will confess that it would be difficult to imagine any other consistent scheme of fundamental principles. And almost all mathematicians will allow to these laws an absolute universality; so that we may apply them without scruple or misgiving, in cases the most remote from those to which our experience has extended. What astronomer would fear to refer to the known laws of motion, in reasoning concerning the double stars; although these objects are at an immeasurably remote distance from that solar system which has been the only field of our observation of mechanical facts? What philosopher, in speculating respecting a magnetic fluid, or a luminiferous ether, would hesitate to apply to it the mechanical principles which are applicable to fluids of known mechanical properties? When we assert that the quantity of motion in the world cannot be increased or diminished by the mutual actions of bodies, does not every mathematician feel convinced that it would be an unphilosophical restriction to limit this proposition to such modes of action as we have tried?

Yet no one can doubt that, in historical fact, these laws were collected from experience. That such is the case, is no matter of conjecture. We know the time, the persons, the circumstances, belonging to each step of each discovery. I have, in the History, given an account of these discoveries; and in the previous chapters of the present work, I have further examined the nature and the import of the principles which were thus brought to light.

Here, then, is an apparent contradiction. Experience, it would seem, has done that which we had proved that she cannot do. She has led men to propositions, universal at least, and to principles which appear to some persons necessary. What is the explanation of this contradiction, the solution of this paradox? Is it true that Experience can reveal to us universal and necessary truths? Does she possess some secret virtue, some unsuspected power, by which she can detect connexions and consequences which we have declared to be out of her sphere? Can she see more than mere appearances, and observe more than mere facts? Can she penetrate, in some way, to the nature of things?—descend below the surface of phenomena to their causes and origins, so as to be able to say what can and what can not be;—what occurrences are partial, and what universal? If this be so, we have indeed mistaken her character and powers; and the whole course of our reasoning becomes precarious and obscure. But, then, when we return upon our path we cannot find the point at which we deviated, we cannot detect the false step in our deduction. It still seems that by experience, strictly so called, we cannot discover necessary and universal truths. Our senses can give us no evidence of a necessary connexion in phenomena. Our observation must be limited, and cannot testify concerning anything which is beyond its limits. A general view of our faculties appears to prove it to be impossible that men should do what the history of the science of mechanics shows that they have done.

2. But in order to try to solve this Paradox, let us again refer to the History of Mechanics. In the cases belonging to that science, in which propositions of the most unquestionable universality, and most approaching to the character of necessary truths, (as, for instance, the laws of motion,) have been arrived at, what is the source of the axiomatic character which the propositions thus assume? The answer to this question will, we may hope, throw some light on the perplexity in which we appear to be involved.

Now the answer to this inquiry is, that the laws of motion borrow their axiomatic character from their being merely *interpretations* of the Axioms of Causation. Those axioms, being exhibitions of the Idea of Cause under various aspects, are of the most rigorous universality and necessity. And so far as the laws of motion are exemplifications of those axioms, these laws must be no less universal and necessary. How these axioms are to be understood;—in what sense *cause* and *effect, action* and *reaction*, are to be taken, experience and observation did, in fact, teach inquirers on this subject; and without this teaching, the laws of motion could never have been distinctly known. If two forces act together, each must produce its effect, by the axiom of causation; and, therefore, the effects of the separate forces must be *compounded*. But a long course of discussion and experiment must instruct men of what kind this *composition* of forces is. Again; action and reaction must be equal; but much thought and some trial were needed to show what *action* and *reaction* are. Those metaphysicians who enunciated Laws of motion without reference to experience, propounded only such laws as were vague and inapplicable. But yet these persons manifested the indestructible conviction, belonging to man's speculative nature, that there exist Laws of motion, that is, universal formulæ, connecting the causes and effects when motion takes place. Those mechanicians, again, who, observed facts involving equilibrium and motion, and stated some narrow rules, without attempting to ascend to any universal and simple principle, obtained laws no less barren and useless than the metaphysicians; for they could not tell in what new cases, or whether in any, their laws would be verified;—they needed a more general rule, to show them the limits of the rule they had discovered. They went wrong in each attempt to solve a new problem, because their interpretation of the terms of the axioms, though true, perhaps, in certain cases, was not right in general.

Thus Pappus erred in attempting to interpret as a case of the lever, the problem of supporting a weight upon an inclined plane; thus Aristotle erred in interpreting the doctrine that the weight of bodies is the cause of their fall; thus Kepler erred in interpreting the rule that the velocity of bodies depends upon the force; thus Bernoulli* erred in interpreting the equality of action and reaction upon a lever in motion. In each of these instances, true doctrines, already established, (whether by experiment or otherwise,) were erroneously applied. And the error was corrected by further reflection, which pointed out that another mode of interpretation was requisite, in order that the axiom which was appealed to in each case might retain its force in the most general sense. And in the reasonings which avoided or corrected such errors, and which led to substantial general truths, the object of the speculator always was to give to the acknowledged maxims which the Idea of Cause suggested, such a signification as should be con-

* *Hist. Ind. Sci.*, B. vi. c. v. sect. 2.

sistent with their universal validity. The rule was not accepted as particular at the outset, and afterwards generalized more and more widely; but from the very first, the universality of the rule was assumed, and the question was, how it should be understood so as to be universally true. At every stage of speculation, the law was regarded as a general law. This was not an aspect which it gradually acquired, by the accumulating contributions of experience, but a feature of its original and native character. *What* should happen universally, experience might be needed to show: but that what happened should happen *universally*, was implied in the nature of knowledge. The universality of the laws of motion was not gathered from experience, however much the laws themselves might be so.

3. Thus we obtain the solution of our Paradox, so far as the case before us is concerned. The laws of motion borrow their *form* from the Idea of Causation, though their *matter* may be given by experience: and hence they possess a universality which experience cannot give. They are certainly and universally valid; and the only question for observation to decide is, how they are to be understood. They are like general mathematical formulæ, which are known to be true, even while we are ignorant what are the unknown quantities which they involve. It must be allowed, on the other hand, that so long as these formulæ are not interpreted by a real study of nature, they are not only useless but prejudicial; filling men's minds with vague general terms, empty maxims, and unintelligible abstractions, which they mistake for knowledge. Of such perversion of the speculative propensities of man's nature, the world has seen too much in all ages. Yet we must not, on that account, despise these forms of truth, since without them, no general knowledge is possible. Without general terms,

and maxims, and abstractions, we can have no science, no speculation; hardly, indeed, consistent thought or the exercise of reason. The course of real knowledge is, to obtain from thought and experience the right interpretation of our general terms, the real import of our maxims, the true generalizations which our abstractions involve.

4. If it be asked, How Experience is able to teach us to interpret aright the general terms which the Axioms of Causation involve;—whence she derives the light which she is to throw on these general notions; the answer is obvious;—namely, that the relations of causation are the *conditions* of Experience;—that the general notions are *exemplified* in the particular cases of which she takes cognizance. The events which take place about us, and which are the objects of our observation, we cannot conceive otherwise than as subject to the laws of cause and effect. Every event must have a cause;—Every effect must be determined by its cause;— these maxims are true of the phenomena which form the materials of our experience. It is precisely to them, that these truths apply. It is in the world which we have before our eyes, that these propositions are universally verified; and it is therefore by the observation of what we see, that we must learn how these propositions are to be understood. Every fact, every experiment, is an example of these statements; and it is therefore by attention to and familiarity with facts and experiments, that we learn the signification of the expressions in which the statements are made; just as in any other case we learn the import of language by observing the manner in which it is applied in known cases. Experience is the interpreter of nature; it being understood that she is to make her interpretation in that comprehensive phraseology which is the genuine language of science.

5. We may return for an instant to the objection, that experience cannot give us general truths, since, after any number of trials confirming a rule, we may, for aught we can foresee, have one which violates the rule. When we have seen a thousand stones fall to the ground, we may see one which does not fall under the same apparent circumstances. How then, it is asked, can experience teach us that *all* stones, rigorously speaking, will fall if unsupported? And to this we reply, that it is not true that we can conceive one stone to be suspended in the air, while a thousand others fall, without believing some peculiar cause to support it; and that, therefore, such a supposition forms no exception to the law, that gravity is a force by which *all* bodies are urged downwards. Undoubtedly we can conceive a body, when dropt or thrown, to move in a line quite different from other bodies: thus a certain missile* used by the natives of Australia, and lately brought to this country, when thrown from the hand in a proper manner, describes a curve, and returns to the place from whence it was thrown. But did any one, therefore, even for an instant suppose that the laws of motion are different for this and for other bodies? On the contrary, was not every person of a speculative turn immediately led to inquire how it was that the known causes which modify motion, the resistance of the air and the other causes, produced in this instance so peculiar an effect? And if the motion had been still more unaccountable, it would not have occasioned any uncertainty whether it were consistent with the agency of gravity and the laws of motion. If a body suddenly alter its direction, or move in any other unexpected manner, we never doubt that there is a cause of the change. We may continue quite ignorant of the nature of this cause, but this ignorance

* Called the Bo-me-rang.

never occasions a moment's doubt that the cause exists and is exactly suited to the effect. And thus experience can prove or discover to us general rules, but she can never prove that general rules do not exist. Anomalies, exceptions, unexplained phenomena, may remind us that we have much still to learn, but they can never make us suppose that truths are not universal. We may observe facts that show us we have not fully understood the meaning of our general laws, but we can never find facts which show our laws to have no meaning. Our experience is bound in by the limits of cause and effect, and can give us no information concerning any region where that relation does not prevail. The whole series of external occurrences and objects, through all time and space, exists only, and is conceived only, as subject to this relation; and therefore we endeavour in vain to imagine to ourselves when and where and how exceptions to this relation may occur. The assumption of the connexion of cause and effect is essential to our experience, as the recognition of the maxims which express this connexion is essential to our knowledge.

6. I have thus endeavoured to explain in some measure how, at least in the field of our mechanical knowledge, experience can discover universal truths, though she cannot give them their universality; and how such truths, though borrowing their form from our ideas, cannot be understood except by the actual study of external nature. And thus with regard to the laws of motion, and other fundamental principles of Mechanics, the analysis of our ideas and the history of the progress of the science well illustrate each other.

If the paradox of the discovery of universal truths by experience be thus solved in one instance, a much wider question offers itself to us;—How far the difficulty, and how far the solution, are applicable to other subjects. It is easy to see that this question involves most grave and extensive doctrines with regard to the whole compass of human knowledge: and the views to which we have been led in the present Book of this work are, we trust, fitted to throw much light upon the general aspect of the subject. But after discussions so abstract, and perhaps obscure, as those in which we have been engaged for some chapters, I willingly postpone to a future occasion an investigation which may perhaps appear to most readers more recondite and difficult still. And we have, in fact, many other special fields of knowledge to survey, before we are led by the order of our subject, to those general questions and doctrines, those antitheses brought into view and again resolved, which a view of the whole territory of human knowledge suggests, and by which the nature and conditions of knowledge are exhibited.

Before we quit the subject of mechanical science we shall make a few remarks on another doctrine which forms part of the established truths of the science, namely, the doctrine of universal gravitation.

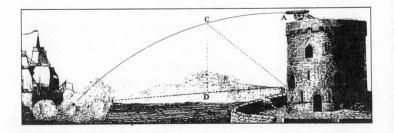

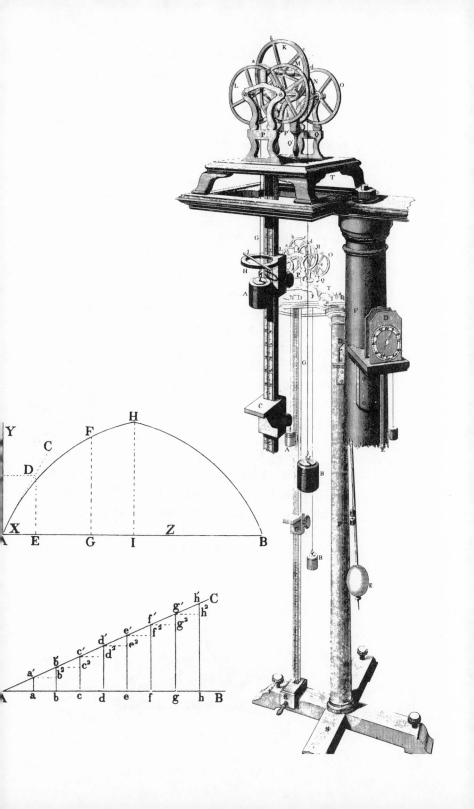

Chapter X.

OF THE GENERAL DIFFUSION OF CLEAR MECHANICAL IDEAS.

1. WE have seen how the progress of knowledge upon the subject of motion and force has produced, in the course of the world's history, a great change in the minds of acute and speculative men; so that such persons can now reason with perfect steadiness and precision upon subjects on which, at first, their thoughts were vague and confused; and can apprehend, as truths of complete certainty and evidence, laws which it required great labour and time to discover. This *complete* developement and clear manifestation of mechanical ideas has taken place only among mathematicians and philosophers. But yet a progress of thought upon such subjects,—an advance from the obscure to the clear, and from errour to truth,—may be traced in the world at large, and among those who have not directly cultivated the exact sciences. This diffused and collateral influence of science manifests itself, although in a wavering and fluctuating manner, by various indications, at various periods of literary history. The opinions and reasonings which are put forth upon mechanical subjects, and above all, the adoption, into common language, of terms and phrases belonging to the prevalent mechanical systems, exhibit to us the most profound discoveries and speculations of philosophers in their effect upon more common

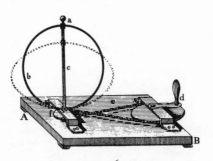

and familiar trains of thought. This effect is by no means unimportant, and we shall point out some examples of such indications as we have mentioned.

2. The discoveries of the ancients in speculative mechanics were, as we have seen, very scanty; and hardly extended their influence to the unmathematical world. Yet the familiar use of the term "center of of gravity" preserved and suggested the most important part of what the Greeks had to teach. The other phrases which they employed, as *momentum, energy, virtue, force*, and the like, never had any exact meaning, even among mathematicians; and therefore never, in the ancient world, became the means of suggesting just habits of thought. I have pointed out, in the History of Science, several circumstances which appear to denote the general confusion of ideas which prevailed upon mechanical subjects during the times of the Roman empire. I have there taken as one of the examples of this confusion, the fable narrated by Pliny and others concerning the echineïs, a small fish, which was said to stop a ship merely by sticking to it[*]. This story was adduced as betraying the absence of any steady apprehension of the equality of action and reaction; since the fish, except it had some immoveable obstacle to hold by, must be pulled forward by the ship, as much as it pulled the ship backward. If the writers who speak of this wonder had shown any perception of the necessity of a reaction, either produced by the rapid motion of the fish's fins in the water, or in any other way, they would not be chargeable with this confusion of thought; but from their expressions it is, I think, evident that they saw no such necessity[†]. Their idea of mechanical action

[*] *Hist. Ind. Sci.* B. iv. c. i. sect. 2.

[†] See Prof. Powell, *On the Nature and Evidence of the Laws of Motion. Reports of the Ashmolean Society.* Oxford. 1837. Professor

was not sufficiently distinct to enable them to see the absurdity of supposing an intense pressure with no obstacle for it to exert itself against.

3. We may trace, in more modern times also, indications of a general ignorance of mechanical truths. Thus the phrase of shooting at an object " point-blank," implies the belief that a cannon-ball describes a path of which the first portion is a straight line. This error was corrected by the true mechanical principles which Galileo and his followers brought to light; but these principles made their way to popular notice, principally in consequence of their application to the motions of the solar system, and to the controversies which took place respecting those motions. Thus by far the most powerful argument against the reception of the Copernican system of the universe, was that of those who asked, Why a stone dropt from a tower was not left behind by the motion of the earth? The answer to this question, now universally familiar, involves a reference to the true doctrine of the composition of motions. Again; Kepler's persevering and strenuous attempts* to frame a physical theory of the universe were frustrated by his ignorance of the first law of motion, which informs us that a body will retain its velocity without any maintaining force. He proceeded upon the supposition that the sun's force was requisite to *keep up* the motion of the planets,

---

Powell has made an objection to my use of this instance of confusion of thought; the remark in the text seems to me to justify what I said in the History. As an evidence that the fish was not supposed to produce its effect by its muscular power acting on the water, we may take what Pliny says, *Nat. Hist.,* xxxii. 1, " Domat mundi rabiem, nullo suo labore; non retinendo, aut alio modo quam adhærendo :" and also what he states in another place (ix. 41,) that when it is preserved in pickle, it may be used in recovering gold which has fallen into a deep well. All this implies adhesion alone, with no conception of reaction.

* *Hist. Ind. Sci.,* B. v. c. iv., and B. VII. c. i.

as well as to deflect and modify it; and he was thus led to a system which represented the sun as carrying round the planets in their orbits by means of a *vortex*, produced by his revolution. The same neglect of the laws of motion presided in the formation of Descartes' system of vortices. Although Descartes had enunciated in words the laws of motion, he and his followers showed that they had not the practical habit of referring to these mechanical principles; and dared not trust the planets to move in free space without some surrounding machinery to support them*.

4. When at last mathematicians, following Newton, had ventured to consider the motion of each planet as a mechanical problem not different in its nature from the motion of a stone cast from the hand; and when the solution of this problem and its immense consequences had become matters of general notoriety and interest; the new views introduced, as is usual, new terms, which soon became extensively current. We meet with such phrases as "flying off in the tangent," and "deflexion from the tangent;" with antitheses between "centripetal" and "centrifugal force," or between "projectile" and "central force." "Centers of force," "disturbing forces," "perturbations," and "perturbations of higher orders," are not unfrequently spoken of: and the expression "to gravitate," and the term "universal gravitation," acquired a permanent place in the language.

Yet for a long time, and even up to the present day, we find many indications that false and confused apprehensions on such subjects are by no means extirpated.

* I have, in the History, applied to Descartes the character which Bacon gives to Aristotle, "Audax simul et pavidus :" though he was bold enough to enunciate the laws of motion without knowing them aright, he had not the courage to leave the planets to describe their orbits by the agency of those laws, without the machinery of contact.

Arguments are urged against the mechanical system of the universe, implying in the opponents an absence of all clear mechanical notions. Many of this class of writers retrograde to Kepler's point of view. This is, for example, the case with Lord Monboddo, who, arguing on the assumption that force is requisite to maintain, as well as to deflect motion, produced a series of attacks upon the Newtonian philosophy; which he inserted in his *Ancient Metaphysics*, published in 1779 and the succeeding years. This writer (like Kepler), measures force by the velocity which the body *has**, not by that which its *gains*. Such a use of language would prevent our obtaining any laws of motion at all. Accordingly, the author, in the very next page to that which I have just quoted, abandons this measure of force, and, in curvilinear motion, measures force by "the fall from the extremity of the arc." Again; in his objections to the received theory, he denies that curvilinear motion is compounded, although his own mode of considering such motion assumes this composition in the only way in which it was ever intended by mathematicians. Many more instances might be adduced to show that a want of cultivation of the mechanical ideas rendered this philosopher incapable of judging of a mechanical system.

The following extract from the *Ancient Metaphysics*, may be sufficient to show the value of the author's criticism on the subjects of which we are now speaking. His object is to prove that there do not exist a centripetal and a centrifugal force in the case of elliptical motion. "Let any man move in a circular or elliptical line described to him; and he will find no tendency in himself either to the center or from it, much less both. If indeed he attempt to make the motion with great velocity, or if he do it carelessly and inattentively, he

* *Anc. Met.* Vol ɪɪ. B. v. c. vi., p. 413.

may go out of the line, either towards the center or from it: but this is to be ascribed, not to the nature of the motion, but to our infirmity; or perhaps to the animal form, which is more fitted for progressive motion in a right line than for any kind of curvilinear motion. But this is not the case with a sphere or spheroid, which is equally adapted to motion in all directions*." We need hardly remind the reader that the manner in which a man running round a small circle, finds it necessary to lean inwards, in order that there may be a centripetal inclination to counteract the centrifugal force, is a standard example of our mechanical doctrines; and this fact (quite familiar in practice as well as theory,) is in direct contradiction of Lord Monboddo's assertion.

5. A similar absence of distinct mechanical thought appears in some of the most celebrated metaphysicians of Germany. I have elsewhere noted† the opinion expressed by Hegel, that the glory which belongs to Kepler has been unjustly transferred to Newton; and I have suggested, as the explanation of this mode of thinking, that Hegel himself, in the knowledge of mechanical truth, had not advanced beyond Kepler's point of view. Persons who possess conceptions of space and number, but who have not learnt to deal with ideas of force and causation, may see more value in the discoveries of Kepler than in those of Newton. Another exemplification of this state of mind may be found in Mr. Schelling's speculations; for instance, in his *Lectures on the Method of Academical Study.* In the twelfth Lecture, on the Study of Physics and Chemistry, he says, (p. 266,) "What the mathematical natural philosophy has done for the knowledge of the laws of the universe since the time that they were discovered by his (Kepler's) godlike genius, is,

* *Anc. Met.*, Vol. i. B. ii. c. 19, p. 264.
† *Hist. Ind. Sci.*, B. vii. c. ii. sect. 5.

181

as is well known, this: it has attempted a construction of those laws which, according to its foundations, is altogether empirical. We may assume it as a general rule, that in any proposed construction, that which is not a pure general form cannot have any scientific import or truth. The foundation from which the centrifugal motion of the bodies of the world is derived, is no necessary form, it is an empirical fact. The Newtonian attractive force, even if it be a necessary assumption for a merely reflective view of the subject, is still of no significance for the Reason, which recognizes only absolute relations. The grounds of the Keplerian laws can be derived, without any empirical appendage, purely from the doctrine of Ideas, and of the two Unities, which are in themseves one Unity, and in virtue of which each being, while it is absolute in itself, is at the same time in the absolute, and reciprocally."

It will be observed, that in this passage our mechanical laws are objected to because they are not necessary results of our ideas; which, however, as we have seen, according to the opinion of some eminent mechanical philosophers, they are. But to assume this evident necessity as a condition of every advance in science, is to mistake the last, perhaps unattainable step, for the first, which lies before our feet. And, without inquiring further about "the Doctrine of the two Unities," or the manner in which from that doctrine we may deduce the Keplerian laws, we may be well convinced that such a doctrine cannot supply any sufficient reason to induce us to quit the inductive path by which all scientific truth up to the present time has been acquired.

6. But without going to schools of philosophy opposed to the Inductive School, we may find many loose and vague habits of thinking on mechanical subjects among the common classes of readers and reasoners. And

there are some familiar modes of employing the phraseology of mechanical science, which are, in a certain
degree, chargeable with inaccuracy, and may produce
or perpetuate confusion. Among such cases we may
mention the way in which the centripetal and centrifugal forces, and also the projectile and central forces
of the planets, are often compared or opposed. Such
antitheses sometimes proceed upon the false notion that
the two members of these pairs of forces are of the
same kind : whereas on the contrary the *projectile* force
is a hypothetical impulsive force which may, at some
former period, have caused the motion to begin ; while
the *central* force is an actual force, which must act continuously and during the whole time of the motion, in
order that the motion may go on in the curve. In the
same manner the *centrifugal* force is not a distinct force
in a strict sense, but only a certain result of the first
law of motion, measured by the portion of *centripetal*
force which counteracts it. Comparisons of quantities
so heterogeneous imply confusion of thought, and often
suggest baseless speculations and imagined reforms of
the received opinions.

7. I might point out other terms and maxims, in
addition to those already mentioned, which, though formerly employed in a loose and vague manner, are now
accurately understood and employed by all just thinkers;
and thus secure and diffuse a right understanding of mechanical truths. Such are *momentum, inertia, quantity
of matter, quantity of motion ;* that *force is proportional
to its effects;* that *action and reaction are equal;* that
what is gained in force by machinery is lost in time ;
that *the quantity of motion in the world cannot be either
increased or diminished.* When the expression of the
truth thus becomes easy and simple, clear and convincing, the meanings given to words and phrases by

discoverers glide into the habitual texture of men's reasonings, and the effect of the establishment of true mechanical principles is felt far from the school of the mechanician. If these terms and maxims are understood with tolerable clearness, they carry the influence of truth to those who have no direct access to its sources. Many an extravagant project in practical machinery, and many a wild hypothesis in speculative physics, has been repressed by the general currency of such maxims as we have just quoted.

8. Indeed so familiar and evident are the elementary truths of mechanics when expressed in this simple form, that they are received as truisms; and men are disposed to look back with surprize and scorn at the speculations which were carried on in neglect of them. The most superficial reasoner of modern times thinks himself entitled to speak with contempt and ridicule of Kepler's hypothesis concerning the physical causes of the celestial motions: and gives himself credit for intellectual superiority, because he sees, as self-evident, what such a man could not discover at all. It is well for such a person to recollect, that the real cause of his superior insight is not the pre-eminence of his faculties, but the successful labours of those who have preceded him. The language which he has learnt to use unconsciously, has been adapted to, and moulded on, ascertained truths. When he talks familiarly of "accelerating forces" and "deflexions from the tangent," he is assuming that which Kepler did not know, and which it cost Galileo and his disciples so much labour and thought to establish. Language is often called an instrument of thought; but it is also the nutriment of thought; or rather, it is the atmosphere in which thought lives: a medium essential to the activity of our speculative power, although invisible and imperceptible in its operation; and an element

modifying, by its qualities and changes, the growth and complexion of the faculties which it feeds. In this way the influence of preceding discoveries upon subsequent ones, of the past upon the present, is most penetrating and universal, though most subtle and difficult to trace. The most familiar words and phrases are connected by imperceptible ties with the reasonings and discoveries of former men and distant times. Their knowledge is an inseparable part of ours; the present generation inherits and uses the scientific wealth of all the past. And this is the fortune, not only of the great and rich in the intellectual world: of those who have the key to the ancient storehouses, and who have accumulated treasures of their own;—but the humblest inquirer, while he puts his reasonings into words, benefits by the labours of the greatest discoverers. When he counts his little wealth, he finds that he has in his hands coins which bear the image and superscription of ancient and modern intellectual dynasties; and that in virtue of this possession, acquisitions are in his power, solid knowledge within his reach, which none could ever have attained to, if it were not that the gold of truth, once dug out of the mine, circulates more and more widely among mankind.

9. Having so fully examined, in the preceding instances, the nature of the progress of thought which science implies, both among the peculiar cultivators of science, and in that wider world of general culture which receives only an indirect influence from scientific discoveries, we shall not find it necessary to go into the same extent of detail with regard to the other provinces of human knowledge. In the case of the Mechanical Sciences, we have endeavoured to show, not only that Ideas are requisite in order to form into a science the Facts which nature offers to us, but that we can advance,

almost or quite, to a complete identification of the Facts with the Ideas. In the sciences to which we now proceed, we shall not seek to fill up the chasm by which Facts and Ideas are separated; but we shall endeavour to detect the Ideas which our knowledge involves, to show how essential these are; and in some respects to trace the mode in which they have been gradually developed among men.

10. The motions of the heavenly bodies, their laws, their causes, are among the subjects of the first division of the Mechanical Sciences; and of these sciences we formerly sketched the history, and have now endeavoured to exhibit the philosophy. If we were to take any other class of motions, *their* laws and causes might give rise to sciences which would be mechanical sciences in exactly the same sense in which Physical Astronomy is so. The phenomena of magnets, of electrical bodies, of galvanical apparatus, seem to form obvious materials for such sciences; and if they were so treated, the philosophy of such branches of knowledge would naturally come under our consideration at this point of our progress.

But on looking more attentively at the sciences of Electricity, Magnetism, and Galvanism, we discover cogent reasons for transferring them to another part of our arrangement; we find it advisable to associate them with Chemistry, and to discuss their principles when we can connect them with the principles of chemical science. For though the first steps and narrower generalizations of these sciences depend upon mechanical ideas, the highest laws and widest generalizations which we can reach respecting them, involve chemical relations. The progress of these portions of knowledge is in some respects opposite to the progress of Physical Astronomy. In this, we begin with phenomena which appear to indicate peculiar and various qualities in the bodies which we consider, (namely, the heavenly bodies,) and we find in the end that all these qualities resolve themselves into one common mechanical property, which exists alike in all bodies and parts of bodies. On the contrary, in studying magnetical and electrical laws, we appear at first to have a single extensive phenomenon, attraction and repulsion: but in our attempts to generalize this phenomenon, we find that it is governed by conditions depending upon something quite separate from the bodies themselves, upon the presence and distribution of peculiar and transitory agencies; and, so far as we can discover, the general laws of these agencies are of a *chemical* nature, and are brought into action by peculiar properties of special substances. In cosmical phenomena, everything, in proportion as it is referred to mechanical principles, tends to simplicity,—to permanent uniform forces,—to one common, positive, property. In magnetical and electrical appearances, on the contrary, the application of mechanical principles leads only to a new complexity, which requires a new explanation; and this explanation involves changeable and various forces,—gradations and oppositions of qualities. The doctrine of the universal gravitation of matter is a simple and ultimate truth, in which the mind can acquiesce and repose. We rank gravity among the mechanical attributes of matter, and we see no necessity to derive it from any ulterior properties. Gravity belongs to matter, independent of any conditions. But the *conditions* of magnetic or electrical activity require investigation as much as the *laws* of their action. Of these conditions no mere mechanical explanation can be given; we are compelled to take along with us chemical properties and relations also: and thus magnetism, electricity, galvanism, are *mechanico-chemical sciences.*

11. Before considering these, therefore, I shall treat

of what I shall call *Secondary Mechanical Sciences;* by which expression I mean the sciences depending upon certain qualities which our senses discover to us in bodies;—*Optics,* which has visible phenomena for its subject; *Acoustics,* the science of hearing; the doctrine of *Heat,* a quality which our touch recognizes: to this last science I shall take the liberty of sometimes giving the name *Thermotics,* analogous to the names of the other two. If our knowledge of the phenomena of Smell and Taste had been successfully cultivated and systematizéd, the present part of our work would be the place for the philosophical discussion of those sensations as the subjects of science.

The branches of knowledge thus grouped in one class involve common Fundamental Ideas, from which their principles are derived in a mode analogous, at least in a certain degree, to the mode in which the principles of the mechanical sciences are derived from the fundamental ideas of causation and reaction. We proceed now to consider these Fundamental Ideas, their nature, development, and consequences.

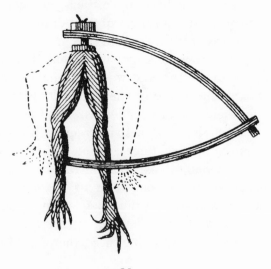

CHAPTER V.

THE ATOMIC THEORY.

1. *The Atomic Theory considered on Chemical Grounds.*—WE have already seen that the combinations which result from chemical affinity are definite, a certain quantity of one ingredient uniting, not with an uncertain, but with a certain quantity of another ingredient. But it was found, in addition to this principle, that one ingredient would often unite with another in different proportions, and that, in such cases, these proportions are multiples one of another. In the three salts formed by potassa with oxalic acid, the quantities of acid which combine with the same quantity of alkali are exactly in the proportion of the numbers 1, 2, 4. And the same rule of the existence of multiple proportions is found to obtain in other cases.

It is obvious that such results will be accounted for, if we suppose the base and the acid to consist each of definite equal particles, and that the formation of the salts above mentioned consists in the combination of one particle of the base with one particle of acid, with two particles of acid, and with four particles of acid, respectively. But further; as we have already stated, chemical affinity is not only definite, but reciprocal. The proportions of potassa and soda which form neutral salts being 590 and 391 in one case, they are so in all cases. These numbers represent the *proportions* of weight in which the two bases, potassa and soda, enter into analogous combinations; 590 of potassa is *equivalent* to 391 of soda. These facts with regard to combination are still expressed by the above supposition of equal particles, assuming that the weights of a particle of potassa and of soda are in the proportion of 590 to 391.

But we pursue our analysis further. We find that potassa is a compound of a metallic base, potassium, and of oxygen, in the proportion of 490 to 100; we suppose, then, that the particle of potassa consists of a particle of potassium and a particle of oxygen, and these latter particles, since we see no present need to suppose them divided, potassium and oxygen being simple bodies, we may call *atoms*, and assume to be indivisible. And by supposing all simple bodies to consist of such atoms, and compounds to be formed by the union of two, or three, or more of such atoms, we explain the occurrence of definite and multiple proportions, and we construct the Atomic Theory.

2. *Hypothesis of Atoms.*—So far as the assumption of such atoms as we have spoken of serves to express those laws of chemical composition which we have referred to, it is a clear and useful generalization. But if the Atomic Theory be put forwards (and its author, Dr. Dalton, appears to have put it forwards with such an intention,) as asserting that chemical elements are really composed of *atoms*, that is, of such particles not further divisible, we cannot avoid remarking, that for such a conclusion, chemical research has not afforded, nor can afford, any satisfactory evidence whatever. The smallest observable quantities of ingredients, as well as the largest, combine according to the laws of proportions

and equivalence which have been cited above. How are we to deduce from such facts any inference with regard to the existence of certain smallest possible particles? The Theory, when dogmatically taught as a physical truth, asserts that all observable quantities of elements *are* composed of proportional numbers of particles which can no further be subdivided; but all which observation teaches us is, that *if* there be such particles, they are smaller than the smallest observable quantities. In chemical experiment, at least, there is not the slightest positive evidence for the existence of such atoms. The assumption of *indivisible* particles, smaller than the smallest observable, which combine, particle with particle, will explain the phenomena; but the assumption of particles bearing this proportion, but *not* possessing the property of indivisibility, will explain the phenomena at least equally well. The decision of the question, therefore, whether the Atomic Hypothesis be the proper way of conceiving the chemical combinations of substances, must depend, not upon chemical facts, but upon our conception of substance. In this sense the question is an ancient and curious controversy, and we shall hereafter have to make some remarks upon it.

3. *Chemical Difficulties of the Hypothesis.*—But before doing this, we may observe that there is no small difficulty in reconciling this hypothesis with the facts of chemistry. According to the theory, all salts, compounded of an acid and a base, are analogous in their atomic constitution; and the number of atoms in one such compound being known or assumed, the number of atoms in other salts may be determined. But when we proceed in this course of reasoning to other bodies, as metals, we find ourselves involved in difficulties. The protoxide of iron is a base which, according to all analogy, must consist of one atom of iron and one of oxygen:

but the peroxide of iron is also a base, and it appears by the analysis of this substance that it must consist of *two-thirds* of an atom of iron and one atom of oxygen. Here, then, our indivisible atoms must be divisible, even upon chemical grounds. And if we attempt to evade this difficulty by making the peroxide of iron consist of two atoms of iron and three of oxygen, we have to make a corresponding alteration in the theoretical constitution of all bodies analogous to the protoxide; and thus we overturn the very foundation of the theory. Chemical facts, therefore, not only do not prove the Atomic Theory as a physical truth, but they are not, according to any modification yet devised of the theory, reconcileable with its scheme.

Nearly the same conclusions result from the attempts to employ the Atomic Hypothesis in expressing another important chemical law;—the law of the combinations of gases according to definite proportions of their volumes, experimentally established by Gay Lussac*. In order to account for this law, it has been very plausibly suggested that all gases, under the same pressure, contain an equal number of atoms in the same space; and that when they combine, they unite atom to atom. Thus one volume of chlorine unites with one volume of hydrogen, and form hydrochloric acid†. But then this hydrochloric acid occupies the space of the two volumes; and therefore the proper number of particles cannot be supplied, and the uniform distribution of atoms in all gases maintained, without dividing into two each of the compound particles, constituted of an atom of chlorine and an atom of hydrogen. And thus in this case, also, the Atomic Theory becomes untenable if it be understood to imply the indivisibility of the atoms.

In all these attempts to obtain a distinct physical

* *Hist. Ind. Sc.*, B. xiv. c. 8.   † Dumas, *Phil. Chim.* 263.

conception of chemical union by the aid of the Atomic Hypothesis, the atoms are conceived to be associated by certain forces of the nature of mechanical attractions. But we have already seen * that no such mode of conception can at all explain or express the facts of chemical combination; and therefore it is not wonderful that when the Atomic Theory attempts to give an account of chemical relations by contemplating them under such an aspect, the facts on which it grounds itself should be found not to authorize its positive doctrines; and that when these doctrines are tried upon the general range of chemical observation, they should prove incapable of even expressing, without self-contradiction, the laws of phenomena.

4. *Grounds of the Atomic Doctrine.*—Yet the doctrine of atoms, or of substance as composed of indivisible particles, has in all ages had great hold upon the minds of physical speculators; nor would this doctrine ever have suggested itself so readily, or have been maintained so tenaciously, as the true mode of conceiving chemical combinations, if it had not been already familiar to the minds of those who endeavour to obtain a general view of the constitution of nature. The grounds of the assumption of the atomic structure of substance are to be found rather in the idea of substance itself, than in the experimental laws of chemical affinity. And the question of the existence of atoms, thus depending upon an idea which has been the subject of contemplation from the very infancy of philosophy, has been discussed in all ages with interest and ingenuity. On this very account it is unlikely that the question, so far as it bears upon chemistry, should admit of any clear and final solution. Still it will be instructive to look back at some of the opinions which have been delivered respecting this doctrine.

* See Chapter I. of this Book.

8. *Arguments for and against Atoms.*—The leading arguments on the two sides of the question, in their most general form, may be stated as follows:—

For the Atomic Doctrine.—The appearances which nature presents are compounded of many parts, but if we go on resolving the larger parts into smaller, and so on successively, we must at last come to something simple. For that which is compound can be so no otherwise than by composition of what is simple; and if we suppose all composition to be removed, which hypothetically we may do, there can remain nothing but a number of simple substances, capable of composition, but themselves not compounded. That is, matter being dissolved, resolves itself into atoms.

Against the Atomic Doctrine.—Space is divisible without limit, as may be proved by geometry; and matter occupies space, therefore matter is divisible without limit, and no portion of matter is *indivisible*, or an *atom*.

And to the argument on the other side just stated, it is replied that we cannot even hypothetically divest a body of composition, if by composition we mean the relation of point to point in space. However small be a particle, it is compounded of parts having relation in space.

The Atomists urge again, that if matter be infinitely divisible, a finite body consists of an infinite number of parts, which is a contradiction. To this it is replied, that the finite body consists of an infinite number of parts in the same sense in which the parts are infinitely small, which is no contradiction.

But the opponents of the Atomists not only rebut, but retort this argument drawn from the notion of infinity. Your atoms, they say, are indivisible by any finite force; therefore they are infinitely hard; and thus your finite particles possess infinite properties. To this the Atomists are wont to reply, that they do not mean the hardness of their particles to be infinite, but only so great as to resist all usual natural forces. But here it is plain that their position becomes untenable; for, in the first place, their assumption of this precise degree of hardness in the particles is altogether gratuitous; and in the next place, if it were granted, such particles are not atoms, since in the next moment the forces of nature may be augmented so as to divide the particle, though hitherto undivided.

Such are the arguments for and against the Atomic Theory in its original form. But when these atoms are conceived, as they have been by Newton, and commonly by his followers, to be solid, hard particles exerting attractive and repulsive forces, a new set of arguments come into play. Of these, the principal one may be thus stated: According to the Atomic Theory thus modified, the properties of bodies depend upon the attractions and repulsions of the particles. Therefore, among other properties of bodies, their hardness depends upon such forces. But if the hardness *of the bodies* depends upon the forces, the repulsion, for instance, of the particles, upon what does the hardness *of the particles* depend? what progress do we make in explaining the properties of bodies, when we assume the same properties in our explanation? and to what purpose do we assume that the particles *are* hard?

BOOK IX.

THE PHILOSOPHY OF BIOLOGY.

CHAPTER I.

ANALOGY OF BIOLOGY WITH OTHER SCIENCES.

1. IN the History of the Sciences, after treating of the Sciences of Classification, we proceeded to what are there termed the Organical Sciences, including in this term Physiology and Comparative Anatomy. A peculiar feature in this group of sciences is that they involve the notion of *living* things. The notion of *Life*, however vague and obscure it may be in men's minds, is apprehended as a peculiar Idea, not resolvable into any other Ideas, such, for instance, as Matter and Motion. The separation between living creatures and inert matter, between organized and unorganized beings, is conceived as a positive and insurmountable barrier. The two classes of objects are considered as of a distinct kind, produced and preserved by different forces. Whether the Idea of Life is really thus original and fundamental, and whether, if so, it be one Idea only, or involve several, it must be the province of true philosophy to determine. What we shall here offer may be considered as an attempt to contribute something to the determination of these questions; but we shall perhaps be able to make it appear that science is at present only in the course of its progress towards a complete solution of such problems.

Since the main feature of those sciences of which we have now to examine the philosophy is, that they involve the Idea of Life, it would be desirable to have them designated by a name expressive of that circumstance. The word *Physiology*, by which they have most commonly been described, means *the Science of Nature*; and though it would be easy to explain, by reference to history, the train of thought by which the word was latterly restricted to *Living Nature*, it is plain that the name is, etymologically speaking, loose and improper. The term *Biology*, which means exactly what we wish to express, *the Science of Life*, has often been used, and has of late become not uncommon among good writers. I shall therefore venture to employ it, in most cases, rather than the word *Physiology*.

2. As I have already intimated, one main inquiry belonging to the Philosophy of Biology, is concerning the Fundamental Idea or Ideas which the science involves. If we look back at the course and the results of our disquisitions respecting other sciences in this work, and assume, as we may philosophically do, that there will be some general analogy between those sciences and this, in their developement and progress, we shall be enabled to anticipate in some measure the nature of the view which we shall now have to take. We have seen that in other subjects the Fundamental Ideas on which science depended, and the Conceptions derived from these, were at first vague, obscure, and confused;—that by gradual steps, by a constant union of thought and observation, these conceptions become more and more clear, more and more definite;—and that when they approached complete distinctness and precision, there were made great positive discoveries into which these conceptions entered, and thus the new precision of thought was fixed and perpetuated in some conspicuous and lasting

truths. Thus we have seen how the first confused mechanical conceptions (Force, and the like,) were, from time to time, growing clearer, down to the epoch of Newton;—how true conceptions of Genera and of wider classes, gradually unfolded themselves among the botanists of the sixteenth and seventeenth centuries;—how the idea of Substance became steady enough to govern the theories of chemists only at the epoch of Lavoisier;—how the Idea of Polarity, although often used by physicists and chemists, is even now somewhat vague and indistinct in the minds of the greater part of speculators. In like manner we may expect to find that the Idea of Life, if indeed *that* be the governing Idea of the Science which treats of Living Things, will be found to have been gradually approaching towards a distinct and definite form among the physiologists of all ages up to the present day. And if this be the case, it may not be considered superfluous, with reference to so interesting a subject, if we employ some space in tracing historically the steps of this progress;—the changes by which the originally loose notion of Life, or of Vital Powers, became more nearly an Idea suited to the purposes of science.

3. But we may safely carry this analogy between Biology and other sciences somewhat further. We have seen, in other sciences, that while men in their speculations were thus tending towards a certain peculiar Idea, but before they as yet saw clearly that it was peculiar and independent, they naturally and inevitably clothed their speculations in conceptions borrowed from some other extraneous idea. And the unsatisfactoriness of all such attempts, and the necessary consequence of this, a constant alteration and succession of such inappropriate hypotheses, were indications and aids of the progress which was going on towards a more genuine form

of the science. For instance, we have seen that in chemistry, so long as men refused to recognize a peculiar and distinct kind of power in the *Affinity* which binds together the elements of bodies, they framed to themselves a series of hypotheses, each constructed according to the prevalent ideas of the time, by which they tried to represent the relation of the compound to the ingredients:—first, supposing that the elements bestowed upon the whole qualities *resembling* their own:—then giving up this supposition, and imagining that the properties of the body depended upon the *shape* of the component particles;—then, as their view expanded, assuming that it was not the shape, but the mechanical *forces* of the particles which gave the body its attributes; —and finally acquiescing in, or rather reluctantly admitting, the idea of *Affinity*, conceived as a peculiar power, different not only from material contact, but from any mechanical or dynamical attraction.

Now we cannot but think it very natural, if we find that the history of Biology offers a series of occurrences of the same nature. The notions of Life in general, or of any Vital Functions or Vital Forces in particular, are obviously very loose and vague as they exist in the minds of most men. The discrepancies and controversies respecting the definitions of all such terms, which are found in all works on physiology, afford us abundant evidence that these notions are not, at least not generally, apprehended with complete clearness and steadiness. We shall therefore find approaches and advances, intermediate steps, gradually leading up to the greatest degree of distinctness which has yet been attained. And in those stages of imperfect apprehension in which the notions of Life and of Vital Powers are still too loose and unformed to be applied independently, we may expect to find them supported and embodied by means of hypotheses borrowed from other subjects, and thus, made so distinct and substantial as to supply at least a temporary possibility of scientific reasoning upon the laws of life.

4. For example, if we suppose that men begin to speculate upon the properties of living things, not acknowledging a peculiar Vital Power, but making use successively of the knowledge supplied by the study of other subjects, we may easily imagine a series of hypotheses along which they would pass.

They would probably, first, in this as in other sciences, have their thoughts occupied by vague and *mystical* notions in which material and spiritual agency, natural and supernatural events, were mixed together without discrimination, and without any clear notion at all. But as they acquired a more genuine perception of the nature of knowledge, they would naturally try to explain vital motions and processes by means of such forces as they had learnt the existence of from other sciences. They might first have a *mechanical* hypothesis, in which the mechanical forces of the solids and fluids which compose organized bodies should be referred to, as the most important influences in the process of life. They might then attend to the actions which the fluids exercise in virtue of their affinity, and might thus form a *chemical* theory. When they had proved the insufficience of these hypotheses, borrowed from the powers which matter exhibits in other cases, they might think themselves authorized to assume some peculiar power or agency, still material, and thus they would have the hypothesis of a *vital fluid*. And if they were driven to reject this, they might think that there was no resource but to assume an immaterial principle of life, and thus they would arrive at the doctrine of an animal *soul*.

Now, through the cycle of hypotheses which we have

thus supposed, physiology has actually passed. The conclusions to which the most philosophical minds have been led by a survey of this progress is, that by the failure of all these theories, men have exhausted this path of inquiry, and shown that scientific truth is to be sought in some other manner. But before I proceed further to illustrate this result, it will be proper, as I have already stated, to exhibit historically the various hypotheses which I have described. In doing this I shall principally follow the *History of Medicine* of Sprengel. It is only by taking for my guide a physiologist of acknowledged science and judgment, that I can hope, on such a subject, to avoid errours of detail. I proceed now to give in succession an account of the Mystical, the Iatrochemical, the Iatromathematical, and the Vital-Fluid Schools; and finally of the Psychical School, who hold the Vital Powers to be derived from the Soul (*Psychè*).

BOOK XI.

OF THE CONSTRUCTION OF SCIENCE.

CHAPTER I.

OF TWO PRINCIPAL PROCESSES BY WHICH SCIENCE IS CONSTRUCTED.

To the subject of the present Book all that has preceded is subordinate and preparatory. The First Part of this work treated of Ideas: we now enter upon the Second Part, in which we have to consider the Knowledge which arises from them. It has already been stated that Knowledge requires us to possess both Facts and Ideas;—that every step in our knowledge consists in applying the ideas and conceptions furnished by our minds to the facts which observation and experiment offer to us. When our conceptions are clear and distinct, when our facts are certain and sufficiently numerous, and when the conceptions, being suited to the nature of the facts, are applied to them so as to produce an exact and universal accordance, we attain knowledge of a precise and comprehensive kind, which we may term *Science*. And we apply this term to our knowledge still more decidedly when, facts being thus included in exact and general propositions, such propositions are, in the same manner, included with equal rigour in propositions of a higher degree of generality; and these again in others of a still wider nature, so as to form a large and systematic whole.

But after thus stating, in a general way, the nature of science, and the elements of which it consists, we have been examining with a more close and extensive scrutiny, some of those elements; and we must now return to our main subject, and apply to it the results of our long investigation. We have been exploring the realm of Ideas; we have been passing in review the difficulties in which the workings of our own minds involve us when we would make our conceptions consistent with themselves: and we have endeavoured to get a sight of the true solutions of these difficulties. We have now to inquire how the results of these long and laborious efforts of thought find their due place in the formation of our knowledge. What do we gain by these attempts to make our notions distinct and consistent; and in what manner is the gain of which we thus become possessed, carried to the general treasure-house of our permanent and indestructible knowledge? After all this battling in the world of ideas, all this struggling with the shadowy and changing forms of intellectual perplexity, how do we secure to ourselves the fruits of our warfare, and assure ourselves that we have really pushed forwards the frontier of the empire of Science? It is by such an appropriation that the task which we have had in our hands during the last nine Books of this work, must acquire its real value and true place in our design.

In order to do this, we must reconsider, in a more definite and precise shape, the doctrine which has already been laid down;—that our knowledge consists in applying Ideas to Facts; and that the conditions of real knowledge are that the ideas be distinct and appropriate, and exactly applied to clear and certain facts. The steps by which our knowledge is advanced are those by which one or the other of these two processes is rendered more complete;—by which *conceptions* are *made more clear* in themselves, or by which the conceptions more strictly *bind together the facts.* These two processes may be considered as together constituting the whole formation of our knowledge; and the principles which have been established in the preceding Books, bear principally upon the former of these two operations; —upon the business of elevating our conceptions to the highest possible point of precision and generality. But these two portions of the progress of knowledge are so clearly connected with each other, that we shall consider them in immediate succession. And having now to consider these operations in a more exact and formal manner than it was before possible to do, we shall designate them by certain constant and technical phrases. We shall speak of the two processes by which we arrive at science, as *the Explication of Conceptions* and *the Colligation of Facts:* we shall show how the discussions in which we have been engaged have been necessary in order to promote the former of these offices; and we shall endeavour to point out modes, maxims, and principles by which the second of the two tasks may also be furthered.

Chapter IV.

OF THE COLLIGATION OF FACTS.

1. Facts such as the last Chapter speaks of are, by means of such Conceptions as are described in the preceding Chapter, bound together so as to give rise to those general Propositions of which Science consists. Thus the Facts that the planets revolve about the sun in certain periodic times and at certain distances, are included and connected in Kepler's Law, by means of such Conceptions as the *squares of numbers*, the *cubes of distances*, and the *proportionality* of these quantities. Again the existence of this proportion in the motions of any two planets, forms a set of Facts which may all be combined by means of the Conception of a certain *central accelerating force*, as was proved by Newton. The whole of our physical knowledge consists in the establishment of such propositions; and in all such cases, Facts are bound together by the aid of suitable Conceptions. This part of the formation of our knowledge I have called the *Colligation of Facts*: and we may apply this term to every case in which, by an act of the intellect, we establish a precise connexion among the phenomena which are presented to our senses. The knowledge of such connexions, accumulated and systematized, is Science. On the steps by which science is thus collected from phenomena we shall proceed now to make a few remarks.

2. Science begins with *Common* Observation of facts, in which we are not conscious of any peculiar discipline or habit of thought exercised in observing. Thus the common perceptions of the appearances and recurrences of the celestial luminaries, were the first steps of Astro-

nomy: the obvious cases in which bodies fall or are supported, were the beginning of Mechanics; the familiar aspects of visible things, were the origin of Optics; the usual distinctions of well-known plants, first gave rise to Botany. Facts belonging to such parts of our knowledge are noticed by us, and accumulated in our memories, in the common course of our habits, almost without our being aware that we are observing and collecting facts. Yet such facts may lead to many scientific truths; for instance, in the first stages of Astronomy (as we have shown in the *History*) such facts lead to Methods of Intercalation and Rules of the Recurrence of Eclipses. In succeeding stages of science, more especial attention and preparation on the part of the observer, and a selection of certain *kinds* of facts, becomes necessary; but there is an early period in the progress of knowledge at which man is a physical philosopher, without seeking to be so, or being aware that he is so.

3. But in all stages of the progress, even in that early one of which we have just spoken, it is necessary, in order that the facts may be fit materials of any knowledge, that they should be decomposed into Elementary Facts, and that these should be observed with precision. Thus, in the first infancy of astronomy, the recurrence of phases of the moon, of places of the sun's rising and setting, of planets, of eclipses, was observed to take place at intervals of certain definite numbers of days, and in a certain exact order; and thus it was, that the observations became portions of astronomical science. In other cases, although the facts were equally numerous, and their general aspect equally familiar, they led to no science, because their exact circumstances were not apprehended. A vague and loose mode of looking at facts very easily observable, left men for a long time under the belief that a body, ten times as heavy as another,

falls ten times as fast;—that objects immersed in water are always magnified, without regard to the form of the surface;—that the magnet exerts an irresistible force;—that crystal is always found associated with ice;—and the like. These and many others are examples how blind and careless man can be, even in observation of the plainest and commonest appearances; and they show us that the mere faculties of perception, although constantly exercised upon innumerable objects, may long fail in leading to any exact knowledge.

4. If we further inquire what was the favourable condition through which some special classes of facts were, from the first, fitted to become portions of science, we shall find it to have been principally this;—that these facts were considered with reference to the Ideas of Time, Number, and Space, which are Ideas possessing peculiar definiteness and precision; so that with regard to them, confusion and indistinctness are hardly possible. The interval from new moon to new moon was always a particular number of days: the sun in his yearly course rose and set near to a known succession of distant objects: the moon's path passed among the stars in a certain order:—these are observations in which mistake and obscurity are not likely to occur, if the smallest degree of attention is bestowed upon the task. To count a number is, from the first opening of man's mental faculties, an operation which no science can render more precise. The relations of space are nearest to those of number in obvious and universal evidence. Sciences depending upon these Ideas arise with the first dawn of intellectual civilization. But few of the other Ideas which man employs in the acquisition of knowledge possess this clearness in their common use. The Idea of *Resemblance* may be noticed, as coming next to those of Space and Number in original precision; and the

Idea of *Cause*, in a certain vague and general mode of application, sufficient for the purposes of common life, but not for the ends of science, exercises a very extensive influence over men's thoughts. But the other Ideas on which science depends, with the Conceptions which arise out of them, are not unfolded till a much later period of intellectual progress; and therefore, except in such limited cases as I have noticed, the observations of common spectators and uncultivated nations, however numerous or varied, are of little or no effect in giving rise to Science.

5. Let us now suppose that, besides common everyday perception of facts, we turn our attention to some other occurrences and appearances, with a design of obtaining from them speculative knowledge. This process is more peculiarly called *Observation*, or, when we ourselves occasion the facts, *Experiment*. But the same remark which we have already made, still holds good here. These facts can be of no value, except they are resolved into those exact Conceptions which contain the essential circumstances of the case. They must be determined, not indeed necessarily, as has sometimes been said, "according to Number, Weight, and Measure;" for, as we have endeavoured to show in the preceding Books*, there are many other Conceptions to which phenomena may be subordinated, quite different from these, and yet not at all less definite and precise. But in order that the facts obtained by observation and experiment may be capable of being used in furtherance of our exact and solid knowledge, they must be apprehended and analyzed according to some Conceptions which, applied for this purpose, give distinct and definite results, such as can be steadily taken hold of and reasoned from; that is, the facts must be referred to Clear and Appro-

Books v., vi., vii., viii., ix., x.

priate Ideas, according to the manner in which we have already explained this condition of the derivation of our knowledge. The phenomena of light, when they are such as to indicate sides in the ray, must be referred to the Conception of *polarization;* the phenomena of mixture, when there is an alteration of qualities as well as quantities, must be combined by a Conception of *elementary composition.* And thus, when mere position, and number, and resemblance, will no longer answer the purpose of enabling us to connect the facts, we call in other Ideas, in such cases more efficacious, though less obvious.

6. But how are we, in these cases, to discover such Ideas, and to judge which will be efficacious, in leading to a scientific combination of our experimental data? To this question, we must in the first place answer, that the first and great instrument by which facts, so observed with a view to the formation of exact knowledge, are combined into important and permanent truths, is that peculiar Sagacity which belongs to the genius of a Discoverer; and which, while it supplies those distinct and appropriate Conceptions which lead to its success, cannot be limited by rules, or expressed in definitions. It would be difficult or impossible to describe in words the habits of thought which led Archimedes to refer the conditions of equilibrium on the lever to the Conception of *pressure,* while Aristotle could not see in them anything more than the results of the strangeness of the properties of the circle;—or which impelled Pascal to explain by means of the Conception of the *weight of air,* the facts which his predecessors had connected by the notion of nature's horrour of a vacuum;—or which caused Vitello and Roger Bacon to refer the magnifying power of a convex lens to the bending of the rays of light towards the perpendicular by *refraction,* while

others conceived the effect to result from the matter of medium, with no consideration of its form. These are what are commonly spoken of as felicitous and inexplicable strokes of inventive talent; and such, no doubt, they are. No rules can ensure to us similar success in new cases; or can enable men who do not possess similar endowments, to make like advances in knowledge.

7. Yet still, we may do something in tracing the process by which such discoveries are made ; and this it is here our business to do. We may observe that these, and the like discoveries, are not improperly described as happy *Guesses;* and that Guesses, in these as in other instances, imply various suppositions made, of which some one turns out to be the right one. We may, in such cases, conceive the discoverer as inventing and trying many conjectures, till he finds one which answers the purpose of combining the scattered facts into a single rule. The discovery of general truths from special facts is performed, commonly at least, and more commonly than at first appears, by the use of a series of Suppositions, or *Hypotheses*, which are looked at in quick succession, and of which the one which really leads to truth is rapidly detected, and when caught sight of, firmly held, verified, and followed to its consequences. In the minds of most discoverers, this process of invention, trial, and acceptance or rejection of the hypothesis, goes on so rapidly that we cannot trace it in its successive steps. But in some instances, we can do so ; and we can also see that the other examples of discovery do not differ essentially from these. The same intellectual operations take place in other cases, although this often happens so instantaneously that we lose the trace of the progression. In the discoveries made by Kepler, we have a curious and memorable exhibition of this process in its details. Thanks to his communicative disposi-

tion, we know that he made nineteen hypotheses with regard to the motion of Mars, and calculated the results of each, before he established the true doctrine, that the planet's path is an ellipse. We know, in like manner, that Galileo made wrong suppositions respecting the laws of falling bodies, and Mariotte, concerning the motion of water in a siphon, before they hit upon the correct view of these cases.

8. But it has very often happened in the history of science, that the erroneous hypotheses which preceded the discovery of the truth have been made, not by the discoverer himself, but by his precursors; to whom he thus owed the service, often an important one in such cases, of exhausting the most tempting forms of errour. Thus the various fruitless suppositions by which Kepler endeavoured to discover the law of refraction, led the way to its real detection by Snell; Kepler's numerous imaginations concerning the forces by which the celestial motions are produced,—his " physical reasonings" as he termed them,—were a natural prelude to the truer physical reasonings of Newton. The various hypotheses by which the suspension of vapour in air had been explained, and their failure, left the field open for Dalton with his doctrine of the mechanical mixture of gases. In most cases, if we could truly analyze the operation of the thoughts of those who make, or who endeavour to make discoveries in science, we should find that many more suppositions pass through their minds than those which are expressed in words; many a possible combination of conceptions is formed and soon rejected. There is a constant invention and activity, a perpetual creating and selecting power at work, of which the last results only are exhibited to us. Trains of hypotheses are called up and pass rapidly in review; and the judgment makes its choice from the varied group.

9. It would, however, be a great mistake to suppose that the hypotheses, among which our choice thus lies, are constructed by an enumeration of obvious cases, or by a wanton alteration of relations which occur in some first hypothesis. It may, indeed, sometimes happen that the proposition which is finally established is such as may be formed, by some slight alteration, from those which are justly rejected. Thus Kepler's elliptical theory of Mars's motions, involved relations of lines and angles much of the same nature as his previous false suppositions: and the true law of refraction so much resembles those erroneous ones which Kepler tried, that we cannot help wondering how he chanced to miss it. But it more frequently happens that new truths are brought into view by the application of new Ideas, not by new modifications of old ones. The cause of the properties of the Lever was learnt, not by introducing any new *geometrical* combination of lines and circles, but by referring the properties to genuine *mechanical* Conceptions. When the Motions of the Planets were to be explained, this was done, not by merely improving the previous notions, of cycles of time, but by introducing the new conception of *epicycles* in space. The doctrine of the Four Simple Elements was expelled, not by forming any new scheme of elements which should impart, according to new rules, their sensible qualities to their compounds, but by considering the elements of bodies as *neutralizing* each other. The Fringes of Shadows could not be explained by ascribing new properties to the single rays of light, but were reduced to law by referring them to the *interference* of several rays.

Since the true supposition is thus very frequently something altogether diverse from all the obvious conjectures and combinations, we see here how far we are from being able to reduce discovery to rule, or to give any precepts by which the want of real invention and sagacity shall be supplied. We may warn and encourage these faculties when they exist, but we cannot create them, or make great discoveries when they are absent.

10. The Conceptions which a true theory requires are very often clothed in a *Hypothesis* which connects with them several superfluous and irrelevant circumstances. Thus the Conception of the Polarization of Light was originally represented under the image of *particles* of light having their poles all turned in the same direction. The Laws of Heat may be made out perhaps most conveniently by conceiving Heat to be a *Fluid*. The Attraction of Gravitation might have been successfully applied to the explanation of facts, if Newton had throughout treated Attraction as the result of an *Ether* diffused through space; a supposition which he has noticed as a possibility. The doctrine of Definite and Multiple Proportions may be conveniently expressed by the hypothesis of *Atoms*. In such cases, the Hypothesis may serve at first to facilitate the introduction of a new Conception. Thus a pervading Ether might for a time remove a difficulty, which some persons find considerable, of imagining a body to exert force at a distance. A Particle with Poles is more easily conceived than Polarization in the abstract. And if hypotheses thus employed will really explain the facts by means of a few simple assumptions, the laws so obtained may afterwards be reduced to a simpler form than that in which they were first suggested. The general laws of Heat, of Attraction, of Polarization, of Multiple Proportions, are now certain, whatever image we may form to ourselves of their ultimate causes.

11. In order, then, to discover scientific truths, suppositions consisting either of new Conceptions, or of new Combinations of old ones, are to be made, till we

find one which succeeds in binding together the Facts. But how are we to find this? How is the trial to be made? What is meant by "success" in these cases? To this we reply, that our inquiry must be, whether the Facts have the same relation in the Hypothesis which they have in reality;—whether the results of our suppositions agree with the phenomena which nature presents to us. For this purpose, we must both carefully observe the phenomena, and steadily trace the consequences of our assumptions, till we can bring the two into comparison. The Conceptions which our hypotheses involve, being derived from certain Fundamental Ideas, afford a basis of rigorous reasoning, as we have shown in the Books respecting those Ideas. And the results to which this reasoning leads, will be susceptible of being verified or contradicted by observation of the facts. Thus the Epicyclical Theory of the Moon, once assumed, determined what the moon's place among the stars ought to be at any given time, and could therefore be tested by actually observing the moon's places. The doctrine that musical strings of the same length, stretched with weights of 1, 4, 9, 16, would give the musical intervals of an octave, a fifth, a fourth, in succession, could be put to the trial by any one whose ear was capable of appreciating those intervals: and the inference which follows from this doctrine by numerical reasoning,—that there must be certain imperfections in the concords of every musical scale,—could in like manner be confirmed by trying various modes of *Temperament*. In like manner all received theories in science, up to the present time, have been established by taking up some supposition, and comparing it, directly or by means of its remoter consequences, with the facts it was intended to embrace. Its agreement, under certain cautions and conditions, of which we may hereafter speak, is held to be the evidence of its truth. It answers its genuine purpose, the Colligation of Facts.

12. When we have, in any subject, succeeded in one attempt of this kind, and obtained some true Bond of Unity by which the phenomena are held together, the subject is open to further prosecution; which ulterior process may, for the most part, be conducted in a more formal and technical manner. The first great outline of the subject is drawn; and the finishing of the resemblance of nature demands a more minute pencilling, but perhaps requires less of genius in the master. In the pursuance of this task, rules and precepts may be given, and features and leading circumstances pointed out, of which it may often be useful to the inquirer to be aware.

Before proceeding further, I shall speak of some characteristic marks which belong to such scientific processes as are now the subject of our consideration, and which may sometimes aid us in determining when the task has been rightly executed.

Chapter XI.

FRANCIS BACON.

1. It is a matter of some difficulty to speak of the character and merits of this illustrious man, as regards his place in that philosophical history with which we are here engaged. If we were to content ourselves with estimating him according to the office which, as we have just seen, he claims for himself†, as merely the harbinger and announcer of a sounder method of scientific inquiry than that which was recognized before him, the task would be comparatively easy. For we might select from his writings those passages in which he has delivered opinions and pointed out processes, then novel and strange, but since confirmed by the experience of actual discoverers, and by the judgments of the wisest of suc-

* *De Augm.*, Lib. IV. c. 1.
† And in other passages: thus, "Ego enim buccinator tantum pugnam non ineo." *Nov. Org.*, Lib. IV. c. 1.

ceeding philosophers; and we might pass by, without disrespect, but without notice, maxims and proposals which have not been found available for use;—views so indistinct and vague, that we are even yet unable to pronounce upon their justice;—and boundless anticipations, dictated by the sanguine hopes of a noble and comprehensive intellect. But if we thus reduce the philosophy of Bacon to that portion which the subsequent progress of science has rigorously verified, we shall have to pass over many of those declarations which have excited most notice in his writings, and shall lose sight of many of those striking thoughts which his admirers most love to dwell upon. For he is usually spoken of, at least in this country, as a teacher who not only commenced, but in a great measure completed, the Philosophy of Induction. He is considered, not only as having asserted some general principles, but laid down the special rules of scientific investigation; as not only one of the Founders, but the supreme Legislator of the modern Republic of Science; not only the Hercules who slew the monsters that obstructed the earlier traveller, but the Solon who established a constitution fitted for all future time.

2. Nor is it our purpose to deny that of such praise he deserves a share which, considering the period at which he lived, is truly astonishing. But it is necessary for us in this place to discriminate and select that portion of his system which, bearing upon *physical* science, has since been confirmed by the actual history of science. Many of Bacon's most impressive and captivating passages contemplate the extension of the new methods of discovering truth to intellectual, to moral, to political, as well as to physical science. And how far, and how, the advantages of the inductive method may be secured for those important branches of speculation, it will at

some future time be a highly interesting task to examine. But our plan requires us at present to omit the consideration of these; for our purpose is to learn what the genuine course of the formation of science is, by tracing it in those portions of human knowledge, which, by the confession of all, are most exact, most certain, most complete. Hence we must here deny ourselves the dignity and interest which float about all speculations in which the great moral and political concerns of men are involved. It cannot be doubted that the commanding position which Bacon occupies in men's estimation arises from his proclaiming a reform in philosophy of so comprehensive a nature;—a reform which was to infuse a new spirit into every part of knowledge. Physical Science has tranquilly and noiselessly adopted many of his suggestions; which were, indeed, her own natural impulses, not borrowed from him; and she is too deeply and satisfactorily absorbed in contemplating her results, to talk much about the methods of obtaining them which she has thus instinctively pursued. But the philosophy which deals with mind, with manners, with morals, with polity, is conscious still of much obscurity and perplexity; and would gladly borrow aid from a system in which aid is so confidently promised. The aphorisms and phrases of the *Novum Organon* are far more frequently quoted by metaphysical, ethical, and even theological writers, than they are by the authors of works on physics.

3. Again, even as regards physics, Bacon's fame rests upon something besides the novelty of the maxims which he promulgated. That a revolution in the method of scientific research was going on, all the greatest physical investigators of the sixteenth century were fully aware, as we have shown in the last chapter. But their writings conveyed this conviction to the public at large somewhat slowly. Men of letters, men of the world, men

of rank, did not become familiar with the abstruse works in which these views were published; and above all, they did not, by such occasional glimpses as they took of the state of physical science, become aware of the magnitude and consequences of this change. But Bacon's lofty eloquence, wide learning, comprehensive views, bold pictures of the coming state of things, were fitted to make men turn a far more general and earnest gaze upon the passing change. When a man of his acquirements, of his talents, of his rank and position, of his gravity and caution, poured forth the strongest and loftiest expressions and images which his mind could supply, in order to depict the "Great Instauration" which he announced;—in order to contrast the weakness, the blindness, the ignorance, the wretchedness, under which men had laboured while they followed the long beaten track, with the light, the power, the privileges, which they were to find in the paths to which he pointed;—it was impossible that readers of all classes should not have their attention arrested, their minds stirred, their hopes warmed; and should not listen with wonder and with pleasure to the strains of prophetic eloquence in which so great a subject was presented. And when it was found that the prophecy was verified; when it appeared that an immense change in the methods of scientific research really *had* occurred;—that vast additions to man's knowledge and power had been acquired, in modes like those which had been spoken of; —that further advances might be constantly looked for;—and that a progress, seemingly boundless, was going on in the direction in which the seer had thus pointed;—it was natural that men should hail him as the leader of the revolution; that they should identify him with the event which he was the first to announce; that they should look upon him as the author of that which he had, as they perceived, so soon and so thoroughly comprehended.

4. For we must remark, that although (as we have seen) he was not the only, nor the earliest writer, who declared that the time was come for such a change, he not only proclaimed it more emphatically, but understood it, in its general character, much more exactly, than any of his contemporaries. Among the maxims, suggestions and anticipations which he threw out, there were many of which the wisdom and the novelty were alike striking to his immediate successors;—there are many which even now, from time to time, we find fresh reason to admire, for their acuteness and justice. Bacon stands far above the herd of loose and visionary speculators who, before and about his time, spoke of the establishment of new philosophies. If we must select some one philosopher as the Hero of the revolution in scientific method, beyond all doubt Francis Bacon must occupy the place of honour.

We shall, however, no longer dwell upon these general considerations, but shall proceed to notice some of the more peculiar and characteristic features of Bacon's philosophy; and especially those views, which, occurring for the first time in his writings, have been fully illustrated and confirmed by the subsequent progress of science, and have become a portion of the permanent philosophy of our times.

5. (I.) The first great feature which strikes us in Bacon's philosophical views is that which we have already noticed;—his confident and emphatic announcement of a *New Era* in the progress of science, compared with which the advances of former times were poor and trifling. This was with Bacon no loose and shallow opinion, taken up on light grounds and involving only vague general notions. He had satisfied himself of the justice

of such a view by a laborious course of research and reflection. In 1605, at the age of forty-four, he published his Treatise of the *Advancement of Learning*, in which he takes a comprehensive and spirited survey of the condition of all branches of knowledge which had been cultivated up to that time. This work was composed with a view to that reform of the existing philosophy which Bacon always had before his eyes; and in the Latin edition of his works, forms the First Part of the *Instauratio Magna*. In the Second Part of the Instauratio, the *Novum Organon*, published in 1620, he more explicitly and confidently states his expectations on this subject. He points out how slightly and feebly the examination of nature had been pursued up to his time, and with what scanty fruit. He notes the indications of this in the very limited knowledge of the Greeks who had till then been the teachers of Europe, in the complaints of authors concerning the subtilty and obscurity of the secrets of nature, in the dissensions of sects, in the absence of useful inventions resulting from theory, in the fixed form which the sciences had retained for two thousand years. Nor, he adds*, is this wonderful; for how little of his thought and labour has man bestowed upon science! Out of twenty-five centuries scarce six have been favourable to the progress of knowledge. And even in those favoured times, natural philosophy received the smallest share of man's attention; while the portion so given was marred by controversy and dogmatism; and even those who have bestowed a little thought upon this philosophy, have never made it their main study, but have used it as a passage or drawbridge to serve other objects. And thus, he says, the great Mother of the Sciences is thrust down with indignity to the offices of a handmaid; is made to minister to the labours of

* Lib. i. Aphor. 78 *et seq.*

medicine or mathematics, or to give the first preparatory tinge to the immature minds of youth. For these and similar considerations of the errours of past time, he draws hope for the future, employing the same argument which Demosthenes uses to the Athenians: "That which is worst in the events of the past, is the best as a ground of trust in the future. For if you had done all that became you, and still had been in this condition, your case might be desperate; but since your failure is the result of your own mistakes, there is good hope that, correcting the errour of your course, you may reach a prosperity yet unknown to you."

6. (II.) All Bacon's hope of improvement indeed was placed in an entire *change of the Method* by which science was pursued; and the boldness, and at the same time, (the then existing state of science being considered) the definiteness of his views of the change that was requisite are truly remarkable.

That all knowledge must begin with observation, is one great principle of Bacon's philosophy; but I hardly think it necessary to notice the inculcation of this maxim as one of his main services to the cause of sound knowledge, since it had, as we have seen, been fully insisted upon by others before him, and was growing rapidly into general acceptance without his aid. But if he was not the first to tell men that they must collect their knowledge from observation, he had no rival in his peculiar office of teaching them *how* science must thus be gathered from experience.

It appears to me that by far the most extraordinary parts of Bacon's works are those in which, with extreme earnestness and clearness, he insists upon a *graduated and successive induction*, as opposed to a hasty transit from special facts to the highest generalizations. The nineteenth Axiom of the First Book of the *Novum Organon*

contains a view of the nature of true science most exact
and profound; and, so far as I am aware, at the time
perfectly new. " There are two ways, and can only be
two, of seeking and finding truth. The one, from sense
and particulars, takes a flight to the most general axioms,
and from those principles and their truth, settled once
for all, invents and judges of intermediate axioms. The
other method collects axioms from sense and particulars,
ascending *continuously and by degrees*, so that in the end
it arrives at the most general axioms; this latter way is
the true one, but hitherto untried."

It is to be remarked, that in this passage Bacon employs the term *axioms* to express any propositions collected from facts by induction, and thus fitted to become
the starting-point of deductive reasonings. How far propositions so obtained may approach to the character of
axioms in the more rigorous sense of the term, we have
already in some measure examined; but that question
does not here immediately concern us. The truly remarkable circumstance is to find this recommendation of a
continuous advance from observation, by limited steps,
through successive gradations of generality, given at a
time when speculative men in general had only just
begun to perceive that they must begin their course
from experience in some way or other. How exactly
this description represents the general structure of the
soundest and most comprehensive physical theories, all
persons who have studied the progress of science up to
modern times can bear testimony; but perhaps this
structure of science cannot in any other way be made so
apparent as by those Tables of successive generalizations
in which we have exhibited the history and constitution
of some of the principal physical sciences, in the Chapter
of the preceding Book which treats of the Logic of
Induction. And the view which Bacon thus took of the

true progress of science was not only new, but, so far as I am aware, has never been adequately illustrated up to the present day.

7. It is true, as I observed in the last chapter, that Galileo had been led to see the necessity, not only of proceeding from experience in the pursuit of knowledge, but of proceeding cautiously and gradually; and he had exemplified this rule more than once, when, having made one step in discovery, he held back his foot, for a time, from the next step, however tempting. But Galileo had not reached this wide and commanding view of the successive subordination of many steps, all leading up at last to some wide and simple general truth. In catching sight of this principle, and in ascribing to it its due importance, Bacon's sagacity, so far as I am aware, wrought unassisted and unrivalled.

8. Nor is there any wavering or vagueness in Bacon's assertion of this important truth. He repeats it over and over again; illustrates it by a great number of the most lively metaphors and emphatic expressions. Thus he speaks of the successive *floors* (*tabulata*) of induction; and speaks of each science as a *pyramid** which has observation and experience for its basis. No images can better exhibit the relation of general and particular truths, as our own Inductive Tables may serve to show.

9. (III.) Again; not less remarkable is his contrasting this true Method of Science (while it was almost, as he says, yet untried) with the ancient and *vicious Method*, which began, indeed, with facts of observation, but rushed

---

* *Aug. Sc.*, Lib. III. c. 4. p. 194. So in other places, as *Nov. Org.*, I. Aphorism 104. " De scientiis tum demum bene sperandum est quando per scalam veram et per gradus continuos, et non intermissos aut hiulcos a particularibus ascendetur ad axiomata minora, et deinde ad media, alia aliis superiora, et postremo demum ad generalissima."

at once, and with no gradations, to the most general principles. For this was the course which had been actually followed by all those speculative reformers who had talked so loudly of the necessity of beginning our philosophy from experience. All these men, if they attempted to frame physical doctrines at all, had caught up a few facts of observation, and had erected a universal theory upon the suggestions which these offered. This process of illicit generalization, or, as Bacon terms it, Anticipation of Nature (*anticipatio naturæ*), in opposition to the Interpretation of Nature, he depicts with singular acuteness, in its character and causes. "These two ways," he says* "both begin from sense and particulars; but their discrepancy is immense. The one merely skims over experience and particulars in a cursory transit; the other deals with them in a due and orderly manner. The one, at its very outset, frames certain general abstract principles, but useless; the other gradually rises to those principles which have a real existence in nature."

"The former path," he adds†, "that of illicit and hasty generalization, is one which the intellect follows when abandoned to its own impulse; and this it does from the requisitions of logic. For the mind has a yearning which makes it dart forth to generalities, that it may have something to rest in; and after a little dallying with experience, becomes weary of it; and all these evils are augmented by logic, which requires these generalities to make a show with in its disputations."

"In a sober, patient, grave intellect," he further adds, "the mind, by its own impulse, (and more especially if it be not impeded by the sway of established opinions) attempts in some measure that other and true way, of gradual generalization; but this it does with small profit;

* *Nov. Org.*, I. Aph. 22.　　　　† *Ib.*, Aph. 20.

for the intellect, except it be regulated and aided, is a faculty of unequal operation, and altogether unapt to master the obscurity of things."

The profound and searching wisdom of these remarks appears more and more, as we apply them to the various attempts which men have made to obtain knowledge; when they begin with the contemplation of a few facts, and pursue their speculations, as upon most subjects they have hitherto generally done; for almost all such attempts have led immediately to some process of illicit generalization, which introduces an interminable course of controversy. In the physical sciences, however, we have the further inestimable advantage of seeing the other side of the contrast exemplified: for many of them, as our Inductive Tables show us, have gone on according to the most rigorous conditions of gradual and successive generalization; and in consequence of this circumstance in their constitution, possess, in each part of their structure, a solid truth, which is always ready to stand the severest tests of reasoning and experiment.

We see how justly and clearly Bacon judged concerning the mode in which facts are to be employed in the construction of science. This, indeed, has ever been deemed his great merit: insomuch that many persons appear to apprehend the main substance of his doctrine to reside in the maxim that facts of observation, and such facts alone, are the essential elements of all true science.

10. (IV.) Yet we have endeavoured to establish the doctrine that facts are but one of two ingredients of knowledge both equally necessary;—that *Ideas* are no less indispensable than facts themselves; and that except these be duly unfolded and applied, facts are collected in vain. Has Bacon then neglected this great portion of

his subject? Has he been led by some partiality of view, or some peculiarity of circumstances, to leave this curious and essential element of science in its pristine obscurity? Was he unaware of its interest and importance?

We may reply that Bacon's philosophy, in its effect upon his readers in general, does *not* give due weight or due attention to the ideal element of our knowledge. He is considered as peculiarly and eminently the asserter of the value of experiment and observation. He is always understood to belong to the experiential, as opposed to the ideal school. He is held up in contrast to Plato and others who love to dwell upon that part of knowledge which has its origin in the intellect of man.

11. Nor can it be denied that Bacon has, in the finished part of his *Novum Organum*, put prominently forwards the necessary dependence of all our knowledge upon Experience, and said little of its dependence, equally necessary, upon the Conceptions which the intellect itself supplies. It will appear, however, on a close examination, that he was by no means insensible or careless of this internal element of all connected speculation. He held the balance, with no partial or feeble hand, between phenomena and ideas. He urged the Colligation of Facts, but he was not the less aware of the value of the Explication of Conceptions.

12. This appears plainly from some remarkable Aphorisms in the *Novum Organum*. Thus, in noticing the causes of the little progress then made by science, he states this:—" In the current Notions, all is unsound, whether they be logical or physical. *Substance, quality, action, passion,* even *being,* are not good Conceptions; still less are *heavy, light, dense, rare, moist, dry, generation, corruption, attraction, repulsion, element, matter, form,* and others of that kind; all are fantastical and ill-defined."

And in his attempt to exemplify his own system, he hesi-
tates* in accepting or rejecting the notions of *elementary,
celestial, rare,* as belonging to fire, since, as he says, they
are vague and ill-defined notions (*notiones vagæ nec bene
terminatæ*).   In that part of his work which appears to
be completed, there is not, so far as I have noticed, any
attempt to fix and define any notions thus complained of
as loose and obscure.   But yet such an undertaking appears to have formed part of his plan ; and in the *Abece-
darium Naturæ* †, which consists of the heads of various
portions of his great scheme, marked by letters of the
alphabet, we find the titles of a series of dissertations
" On the Conditions of Beings," which must have had for
their object the elucidation of divers Notions essential to
science, and which would have been contributions to the
Explication of Conceptions, such as we have attempted
in a former part of this work.   Thus some of the subjects
of these dissertations are ;—Of Much and Little ;—Of
Durable and Transitory ;—Of Natural and Monstrous ;—
Of Natural and Artificial.   When the philosopher of
induction came to discuss these, considered as *conditions
of existence,* he could not do other than develope, limit,
methodize, and define the Ideas involved in these Notions,
so as to make them consistent with themselves, and a fit
basis of demonstrative reasoning.   His task would have
been of the same nature as ours has been, in that part
of this work which treats of the Fundamental Ideas of
the various classes of sciences.

13. Thus Bacon, in his speculative philosophy, took
firmly hold of both the handles of science ; and if he had
completed his scheme, would probably have given due
attention to Ideas, no less than to Facts, as an element
of our knowledge ; while in his view of the general

* *Nov. Org.,* Lib. II. Aph. 19.
† *Inst. Mag.,* Par. III. (Vol. VIII. p. 244.)

method of ascending from facts to principles, he displayed a sagacity truly wonderful. But we cannot be surprized, that in attempting to exemplify the method which he recommended, he should have failed. For the method could be exemplified only by some important discovery is physical science; and great discoveries, even with the most perfect methods, do not come at command. Moreover although the general structure of his scheme was correct, the precise import of some of its details could hardly be understood, till the actual progress of science had made men somewhat familiar with the kind of steps which it included.

14. (V.) Accordingly, Bacon's *Inquisition into the Nature of Heat*, which is given in the Second Book of the *Novum Organon* as an example of the mode of interrogating Nature, cannot be looked upon otherwise than as a complete failure. This will be evident if we consider that, although the exact nature of heat is still an obscure and controverted matter, the science of Heat now consists of many important truths; and that to none of these truths is there any approximation in Bacon's essay. From his process he arrives at this, as the "forma or true definition" of heat;—"that it is an expansive, restrained motion, modified in certain ways, and exerted in the smaller particles of the body." But the steps by which the science of Heat really advanced were, (as may be seen in the history* of the subject,) these;—The discovery of a *measure* of heat or temperature (the thermometer); The establishment of the *laws* of conduction and radiation; of the *laws* of specific heat, latent heat, and the like. Such steps have led to Ampère's *hypothesis*†, that heat consists in the vibrations of an imponderable fluid; and to Laplace's *hypothesis*, that temperature consists in the internal radiation of such a fluid. These hypotheses

* *Hist. Ind. Sci.*, B. x. c. i.   † *Ib.*, c. iv.

cannot yet be said to be even probable; but at least they are so modified as to include some of the preceding laws which are firmly established; whereas Bacon's hypothetical motion includes no laws of phenomena, explains no process, and is indeed itself an example of illicit generalization.

15. One main ground of Bacon's ill fortune in this undertaking appears to be, that he was not aware of an important maxim of inductive science, that we must first obtain the *measure* and ascertain the *laws* of phenomena, before we endeavour to discover their *causes*. The whole history of thermotics up to the present time has been occupied with the *former* step, and the task is not yet completed: it is no wonder, therefore, that Bacon failed entirely, when he so prematurely attempted the *second*. His sagacity had taught him that the progress of science must be gradual; but it had not led him to judge adequately how gradual it must be, nor of what different kinds of inquiries, taken in due order, it must needs consist, in order to obtain success.

Another mistake, which could not fail to render it unlikely that Bacon should really exemplify his precepts by any actual advance in science, was, that he did not justly appreciate the sagacity, the inventive genius, which all discovery requires. He conceived that he could supersede the necessity of such peculiar endowments. "Our method of discovery in science," he says*, "is of such a nature, that there is not much left to acuteness and strength of genius, but all degrees of genius and intellect are brought nearly to the same level." And he illustrates this by comparing his method to a pair of compasses, by means of which a person with no manual skill may draw a perfect circle. In the same spirit he speaks of proceeding by *due rejections;* and appears to

* *Nov. Org.*, Lib. i. Aph. 61.

imagine that when we have obtained a collection of facts, if we go on successively rejecting what is false, we shall at last find that we have, left in our hands, that scientific truth which we seek. I need not observe how far this view is removed from the real state of the case. The necessity of a *conception* which must be furnished by the mind in order to bind together the facts, could hardly have escaped the eye of Bacon, if he had cultivated more carefully the ideal side of his own philosophy. And any attempts which he could have made to construct such conceptions by mere rule and method, must have ended in convincing him that nothing but a peculiar inventive talent could supply that which was thus not contained in the facts, and yet was needed for the discovery.

16. (VI.) Since Bacon, with all his acuteness, had not divined circumstances so important in the formation of science, it is not wonderful that his attempt to reduce this process to a *Technical Form* is of little value. In the first place, he says[*], we must prepare a natural and experimental history, good and sufficient; in the next place, the instances thus collected are to be arranged in Tables in some orderly way; and then we must apply a legitimate and true induction. And in his example[†], he first collects a great number of cases in which heat appears under various circumstances, which he calls "a Muster of Instances before the intellect," (*comparentia instantiarum ad intellectum*,) or a *Table of the Presence* of the thing sought. He then adds a *Table of its Absence* in proximate cases, containing instances where heat does not appear; then a *Table of Degrees*, in which it appears with greater or less intensity. He then adds[‡], that we must try to exclude several obvious suppositions, which he does by reference to some of the instances he

---

[*] *Nov. Org.*, Lib. II. Aph. 10.   [†] Aph. 11.
[‡] Aph. 15. p. 105.

has collected; and this step he calls the *Exclusive*, or the *Rejection of Natures*. He then observes, (and justly,) that whereas truth emerges more easily from errour than from confusion, we may, after this preparation, *give play to the intellect*, (fiat permissio intellectus,) and make an attempt at induction, liable afterwards to be corrected; and by this step, which he terms his *First Vindemiation*, or *Inchoate Induction*, he is led to the proposition concerning heat, which we have stated above.

17. In all the details of his example he is unfortunate. By proposing to himself to examine at once into the *nature* of heat, instead of the laws of special classes of phenomena, he makes, as we have said, a fundamental mistake; which is the less surprizing since he had before him so few examples of the right course in the previous history of science. But further, his collection of instances is very loosely brought together; for he includes in his list the *hot* taste of aromatic plants, the *caustic* effects of acids, and many other facts which cannot be ascribed to heat without a studious laxity in the use of the word. And when he comes to that point where he permits his intellect its range, the conception of *motion* upon which it at once fastens, appears to be selected with little choice or skill, the suggestion being taken from flame*, boiling liquids, a blown fire, and some other cases. If from such examples we could imagine heat to be motion, we ought at least to have some gradation to cases of heat where no motion is visible, as in a red-hot iron. It would seem that, after a large collection of instances had been looked at, the intellect, even in its first attempts, ought not to have dwelt upon such an hypothesis as this.

18. After these steps, Bacon speaks of several classes of instances which, singling them out of the general and

* Page 110.

indiscriminate collection of facts, he terms *Instances
with Prerogative;* and these he points out as peculiar
aids and guides to the intellect in its task. These Instances with Prerogative have generally been much
dwelt upon by those who have commented on the *Novum
Organon.* Yet, in reality, such a classification, as has
been observed by one of the ablest writers of the present day[*], is of little service in the task of induction.
For the instances are, for the most part, classed, not
according to the ideas which they involve, or to any
obvious circumstance in the facts of which they consist,
but according to the extent or manner of their influence
upon the inquiry in which they are employed. Thus we
have Solitary Instances, Migrating Instances, Ostensive
Instances, Clandestine Instances, so termed according to
the degree in which they exhibit, or seem to exhibit, the
property whose nature we would examine. We have
Guide-Post Instances, (*Instantiæ Crucis,*) Instances of
the Parted Road, of the Doorway, of the Lamp, according to the guidance they supply to our advance. Such
a classification is much of the same nature as if, having
to teach the art of building, we were to describe tools
with reference to the amount and place of the work
which they must do, instead of pointing out their construction and use:—as if we were to inform the pupil
that we must have tools for lifting a stone up, tools for
moving it sideways, tools for laying it square, tools for
cementing it firmly. Such an enumeration of ends would
convey little instruction as to the means. Moreover,
many of Bacon's classes of instances are vitiated by the
assumption that the "form," that is, the general law and
cause of the property which is the subject of investigation, is to be looked for directly in the instances;

* Herschel, *On the Study of Nat. Phil.*, Art. 192.

which, as we have seen in his inquiry concerning heat, is a fundamental errour.

19. Yet his phraseology in some cases, as in the *instantia crucis,* serves well to mark the place which certain experiments hold in our reasonings: and many of the special examples which he gives are full of acuteness and sagacity. Thus he suggests swinging a pendulum in a mine, in order to determine whether the attraction of the earth arises from the attraction of its parts; and observing the tide at the same moment in different parts of the world, in order to ascertain whether the motion of the water is expansive or progressive; with other ingenious proposals. These marks of genius may serve to counterbalance the unfavourable judgment of Bacon's aptitude for physical science which we are sometimes tempted to form, in consequence of his false views on other points; as his rejection of the Copernican system, and his undervaluing Gilbert's magnetical speculations. Most of these errours arose from a too ambitious habit of intellect, which would not be contented with any except very wide and general truths; and from an indistinctness of mechanical, and perhaps, in general, of mathematical ideas :—defects which Bacon's own philosophy was directed to remedy, and which, in the progress of time, it has remedied in others.

20. (VII.) Having thus freely given our judgment concerning the most exact and definite portion of Bacon's precepts, it cannot be necessary for us to discuss at any length the value of those more vague and general *Warnings* against prejudice and partiality, against intellectual indolence and presumption, with which his works abound. His advice and exhortations of this kind are always expressed with energy and point, often clothed in the happiest forms of imagery; and hence it has come to pass, that such passages are perhaps more familiar to the

general reader than any other parts of his writings. Nor are Bacon's counsels without their importance, when we have to do with those subjects in which prejudice and partiality exercise their peculiar sway. Questions of politics and morals, of manners, taste, or history, cannot be subjected to a scheme of rigorous induction; and though on such matters we venture to assert general principles, these are commonly obtained with some degree of insecurity, and depend upon special habits of thought, not upon mere logical connexion. Here, therefore, the intellect may be perverted, by mixing, with the pure reason, our gregarious affections, or our individual propensities; the false suggestions involved in language, or the imposing delusions of received theories. In these dim and complex labyrinths of human thought, *the Idol of the Tribe*, or *of the Den*, *of the Forum*, or *of the Theatre*, may occupy men's minds with delusive shapes, and may obscure or pervert their vision of truth. But in that Natural Philosophy with which we are here concerned, there is little opportunity for such influences. As far as a physical theory is completed through all the steps of a just induction, there is a clear daylight diffused over it which leaves no lurking-place for prejudice. Each part can be examined separately and repeatedly; and the theory is not to be deemed perfect till it will bear the scrutiny of all sound minds alike. Although, therefore, Bacon, by warning men against the idols or fallacious images above spoken of, may have guarded them from dangerous errour, his precepts have little to do with Natural Philosophy: and we cannot agree with him when he says*, that the doctrine concerning these idols bears the same relation to the interpretation of nature as the doctrine concerning sophistical paralogisms bears to common logic.

* *Nov. Org.*, Lib. I. Aph. 40.

21. (VIII.) There is one very prominent feature in Bacon's speculations which we must not omit to notice; it is a leading and constant object with him to apply his knowledge to *Use*. The insight which he obtains into nature, he would employ in commanding nature for the service of man. He wishes to have not only principles but works. The phrase which best describes the aim of his philosophy is his own*, "Ascendendo ad *axiomata*, descendendo ad *opera*." This disposition appears in the first aphorism of the *Novum Organon*, and runs through the work. "Man, the *minister* and interpreter of nature, *does* and understands, so far as he has, in fact or in thought, observed the course of nature; and he cannot know or *do* more than this." It is not necessary for us to dwell much upon this turn of mind; for the whole of our present inquiry goes upon the supposition that an acquaintance with the laws of nature is worth our having for its own sake. It may be universally true, that Knowledge is Power; but we have to do with it not as Power, but as Knowledge. It is the formation of Science, not of Art, with which we are here concerned. It may give a peculiar interest to the history of science, to show how it constantly tends to provide better and better for the wants and comforts of the body; but *that* is not the interest which engages us in our present inquiry into the nature and course of philosophy. The consideration of the means which promote man's material well-being often appears to be invested with a kind of dignity, by the discovery of general laws which it involves; and the satisfaction which rises in our minds at the contemplation of such cases, men sometimes ascribe, with a false ingenuity, to the love of mere bodily enjoyment. But it is never difficult to see that this baser and coarser element is not the real source of our admiration. Those

* *Nov. Org.*, Lib. I. Ax. 103.

who hold that it is the main business of science to construct instruments for the uses of life, appear sometimes
to be willing to accept the consequence which follows
from such a doctrine, that the first shoemaker was a
philosopher worthy of the highest admiration*. But
those who maintain such paradoxes, often, by a happy
inconsistency, make it their own aim, not to devise some
improved covering for the feet, but to delight the mind
with acute speculations, exhibited in all the graces of
wit and fancy.

It has been said† that the key of the Baconian doctrine consists in two words, Utility and Progress. With
regard to the latter point, we have already seen that the
hope and prospect of a boundless progress in human
knowledge had sprung up in men's minds, even in the
early times of imperial Rome; and were most emphatically expressed by that very Seneca who disdained to
reckon the worth of knowledge by its value in food and
clothing. And when we say that Utility was the great
business of Bacon's philosophy, we forget one-half of his
characteristic phrase. "Ascendendo ad axiomata," no
less than "descendendo ad opera," was, he repeatedly
declared, the scheme of his path. He constantly spoke,
we are told by his secretary‡, of two kinds of experiments, *experimenta fructifera*, and *experimenta lucifera*.

Again; when we are told by modern writers that
Bacon merely recommended such induction as all men
instinctively practise, we ought to recollect his own
earnest and incessant declarations to the contrary. The
induction hitherto practised is, he says, of no use for
obtaining solid science. There are two ways§, "hæc via
in usu est," "altera vera, sed intentata." Men have con-

* *Edinb. Rev.*, No. cxxxii. p. 65.   † *Ib.*
‡ Pref. to the *Nat. Hist.*, i. 243.
§ *Nov. Org.*, Lib. i. Aph. 19.

stantly been employed in *anticipation*; in illicit induction. The intellect left to itself rushes on in this road*; the conclusions so obtained are persuasive+; far more persuasive than inductions made with due caution‡. But still this method must be rejected if we would obtain true knowledge. We shall then at length have ground of good hope for science when we proceed in another manner§. We must rise, not by a leap, but by small steps, by successive advances, by a gradation of ascents, trying our facts, and clearing our notions at every interval. The scheme of true philosophy, according to Bacon, is not obvious and simple, but long and technical, requiring constant care and self-denial to follow it. And we have seen that, in this opinion, his judgment is confirmed by the past history and present condition of science.

Again; it is by no means a just view of Bacon's character to place him in contrast to Plato. Plato's philosophy was the philosophy of Ideas; but it was not left for Bacon to set up the philosophy of Facts in opposition to that of Ideas. That had been done fully by the speculative reformers of the sixteenth century. Bacon had the merit of showing that Facts and Ideas must be combined; and not only so, but of divining many of the special rules and forms of this combination, when as yet there were no examples of them, with a sagacity hitherto quite unparalleled.

22. (IX.) With Bacon's unhappy political life we have here nothing to do. But we cannot but notice with pleasure how faithfully, how perseveringly, how energetically he discharged his great philosophical office of a Reformer of Methods. He had conceived the pur-

---

* *Nov. Org.*, Lib. I. Aph. 20.　　+ Aph. 27.　　‡ *Ib.*, 28.

§ Aph. 104. So Aph. 105. "In constituendo axiomate forma *inductionis* alia quam adhuc in usu fuit excogitanda est," &c.

pose of making this his object at an early period. When meditating the continuation of his *Novum Organon*, and speaking of his reasons for trusting that his work will reach some completeness of effect, he says*, "I am by two arguments thus persuaded. First, I think thus from the zeal and constancy of my mind, which has not waxed old in this design, nor, after so many years, grown cold and indifferent; I remember that about forty years ago I composed a juvenile work about these things, which with great contrivance and a pompous title I called *temporis partum maximum*, or the most considerable birth of time; Next, that on account of its usefulness, it may hope the Divine blessing." In stating the grounds of hope for future progress in the sciences, he says†: "Some hope may, we conceive, be ministered to men by our own example: and this we say, not for the sake of boasting, but because it is useful to be said. If any despond, let them look at me, a man among all others of my age most occupied with civil affairs, nor of very sound health, (which brings a great loss of time;) also in this attempt the first explorer, following the footsteps of no man, nor communicating on these subjects with any mortal; yet, having steadily entered upon the true road and made my mind submit to things themselves, one who has, in this undertaking, made, (as we think,) some progress." He then proceeds to speak of what may be done by the combined and more prosperous labours of others, in that strain of noble hope and confidence, which rises again and again, like a chorus, at intervals in every part of his writings. In the *Advancement of Learning* he had said, "I could not be true and constant to the argument I handle, if I were not willing to go beyond others, but yet not more willing than to have others go beyond me again." In the Preface to the

* *Ep. ad P. Fulgentium. Op.*, x. 330.　　† *Nov. Org.*, i. Aph. 113.

Instauratio Magna, he had placed among his postulates those expressions which have more than once warmed the breast of a philosophical reformer*. "Concerning ourselves we speak not; but as touching the matter which we have in hand, this we ask;—that men be of good hope, neither feign and imagine to themselves this our Reform as something of infinite dimension and beyond the grasp of mortal man, when in truth it is the end and true limit of infinite errour; and is by no means unmindful of the condition of mortality and humanity, not confiding that such a thing can be carried to its perfect close in the space of a single age, but assigning it as a task to a succession of generations." In a later portion of the *Instauratio* he says: "We bear the strongest love to the *human republic,* our common country; and we by no means abandon the hope that there will arise and come forth some man among posterity, who will be able to receive and digest all that is best in what we deliver; and whose care it will be to cultivate and perfect such things. Therefore, by the blessing of the Deity, to tend to this object, to open up the fountains, to discover the useful, to gather guidance for the way, shall be our task; and from this we shall never, while we remain in life, desist."

23. (X.) We may add, that the spirit of piety as well as of hope which is seen in this passage, appears to have been habitual to Bacon at all periods of his life. We find in his works several drafts of portions of his great scheme, and several of them begin with a prayer. One of these entitled, in the edition of his works, "The Student's Prayer," appears to me to belong probably to his early youth. Another, entitled "The Writer's Prayer," is inserted at the end of the Preface of the *Instauratio,* as it was finally published. I will conclude my notice of this wonderful man by inserting here these two prayers.

* See the motto to Kant's *Kritik der Reinen Vernunft.*

"To God the Father, God the Word, God the Spirit, we pour forth most humble and hearty supplications; that he, remembering the calamities of mankind, and the pilgrimage of this our life, in which we wear out days few and evil, would please to open to us new refreshments out of the fountains of his goodness for the alleviating of our miseries. This also we humbly and earnestly beg, that human things may not prejudice such as are divine; neither that, from the unlocking of the gates of sense, and the kindling of a greater natural light, anything of incredulity, or intellectual night, may arise in our minds towards divine mysteries. But rather, that by our mind thoroughly cleansed and purged from fancy and vanities, and yet subject and perfectly given up to the Divine oracles, there may be given unto faith the things that are faith's."

"Thou, O Father, who gavest the visible light as the first-born of thy creatures, and didst pour into man the intellectual light as the top and consummation of thy workmanship, be pleased to protect and govern this work, which coming from thy goodness, returneth to thy glory. Thou, after thou hadst reviewed the works which thy hands had made, beheldest that everything was very good, and thou didst rest with complacency in them. But man, reflecting on the works which he had made, saw that all was vanity and vexation of spirit, and could by no means acquiesce in them. Wherefore, if we labour in thy works with the sweat of our brows, thou wilt make us partakers of thy vision and thy Sabbath. We humbly beg that this mind may be steadfastly in us; and that thou, by our hands, and also by the hands of others on whom thou shalt bestow the same spirit, wilt please to convey a largess of new alms to thy family of mankind. These things we commend to thy everlasting love, by our Jesus, thy Christ, God with us. Amen."

Chapter V.

ANALYSIS OF THE PROCESS OF INDUCTION.

Sect. I.—*The Three Steps of Induction.*

1. When facts have been decomposed and phenomena measured, the philosopher endeavours to combine them into general laws, by the aid of ideas and conceptions, these being illustrated and regulated by such means as we have spoken of in the last two chapters. In this task, of gathering laws of nature from observed facts, as we have already said*, the natural sagacity of gifted minds is the power by which the greater part of the successful results have been obtained; and this power will probably always be more efficacious than any Method can be. Still there are certain methods of procedure which may in such investigations give us no inconsiderable aid, and these I shall endeavour to expound.

2. For this purpose, I remark that the Colligation of ascertained facts into general propositions may be considered as containing three steps, which I shall term *the Selection of the Idea, the Construction of the Conception,* and *the Determination of the Magnitudes.* It will be recollected that by the word *Idea,* (or Fundamental Idea,) used in a peculiar sense, I mean certain wide and general fields of intelligible relation, such as Space, Number, Cause, Likeness; while by *Conception* I denote more special modifications of these ideas, as a *circle,* a *square number,* a *uniform force,* a *like form* of flower. Now in

---

* B. xi. c. vi.

244

order to establish any law by reference to facts, we must
select the *true Idea* and the *true Conception.* For example; when Hipparchus found* that the distance of the
bright star Spica Virginis from the equinoxial point had
increased by two degrees in about two hundred years,
and desired to reduce this change to a law, he had first
to assign, if possible, the *idea* on which it depended;—
whether it was regulated for instance, by *space,* or by
time; whether it was determined by the positions of other
stars at each moment, or went on progressively with the
lapse of ages. And when there was found reason to
select *time* as the regulative *idea* of this change, it was
then to be determined how the change went on with the
time;—whether uniformly, or in some other manner:
the *conception,* or the rule of the progression, was to be
rightly constructed. Finally, it being ascertained that
the change did go on uniformly, the question then occurred
what was its *amount:*—whether exactly a degree in a
century, or more, or less, and how much: and thus the
determination of the *magnitude* completed the discovery
of the law of phenomena respecting this star.

3. Steps similar to these three may be discerned in
all other discoveries of laws of nature. Thus, in investigating the laws of the motions of the sun, moon or
planets, we find that these motions may be resolved,
besides a uniform motion, into a series of partial motions,
or Inequalities; and for each of these Inequalities, we
have to learn upon what it directly depends, whether
upon the progress of time only, or upon some configura-
tion of the heavenly bodies in space; then, we have to
ascertain its law; and finally, we have to determine what
is its amount. In the case of such Inequalities, the
fundamental element on which the Inequality depends, is
called the *Argument.* And when the Inequality has been

* *Hist. Ind. Sci.,* B. III. c. iv. sect. 3.

fully reduced to known rules, and expressed in the form of a Table, the Argument is the fundamental series of numbers which stands in the margin of the Table, and by means of which we refer to the other numbers which express the Inequality. Thus, in order to obtain from a Solar Table the Inequality of the sun's annual motion, the Argument is the number which expresses the day of the year; the Inequalities for each day being (in the Table) ranged in a line corresponding to the days. Moreover, the Argument of an Inequality being assumed to be known, we must, in order to calculate the Table, that is, in order to exhibit the law of nature, know also the *Law* of the Inequality, and its *Amount*. And the investigation of these three things, the Argument, the Law, and the Amount of the Inequality, represents the three steps above described, the Selection of the Idea, the Construction of the Conception, and the Determination of the Magnitude.

4. In a great body of cases, *mathematical* language and calculation are used to express the connexion between the general law and the special facts. And when this is done, the three steps above described may be spoken of as the Selection of the *Independent Variable*, the Construction of the *Formula*, and the Determination of the *Coefficients*. It may be worth our while to attend to an exemplification of this. Suppose then, that, in such observations as we have just spoken of, namely, the shifting of a star from its place in the heavens by an unknown law, astronomers had, at the end of three successive years, found that the star had removed by 3, by 8, and by 15 minutes from its original place. Suppose it to be ascertained also, by methods of which we shall hereafter treat, that this change depends upon the time; we must then take the *time*, (which we may denote by the symbol t,) for the *independent variable*. But though the star changes

its place with the time, the change is not proportional to the time; for its motion which is only 3 minutes in the first year, is 5 minutes in the second year, and 7 in the third. But it is not difficult for a person a little versed in mathematics to perceive that the series 3, 8, 15, may be obtained by means of two terms, one of which is proportional to the time, and the other to the square of the time; that is, it is expressed by the *formula at + btt*. The question then occurs, what are the values of the *coefficients* a and b; and a little examination of the case shows us that a must be 2, and b, 1: so that the formula is $2t + tt$. Indeed if we add together the series 2, 4, 6, which expresses the change proportional to the time, and 1, 4, 9, which is proportional to the square of the time, we obtain the series 3, 8, 15, which is the series of numbers given by observation. And thus the three steps which give us the Idea, the Conception, and the Magnitudes; or the Argument, the Law, and the Amount, of the change; give us the Independent Variable, the Formula, and the Coefficients, respectively.

We now proceed to offer some suggestions of methods by which each of these steps may be in some degree promoted.

APHORISMS CONCERNING IDEAS.

I.

MAN is the Interpreter of Nature, Science the right interpretation. (Book I. Chapter 1.)

II.

The *Senses* place before us the *Characters* of the Book of Nature; but these convey no knowledge to us, till we have discovered the Alphabet by which they are to be read. (I. 2.)

III.

The *Alphabet*, by means of which we interpret Phenomena, consists of the *Ideas* existing in our own minds; for these give to the phenomena that coherence and significance which is not an object of sense. (I. 2.)

IV.

The antithesis of *Sense* and *Ideas* is the foundation of the Philosophy of Science. No knowledge can exist without the union, no philosophy without the separation, of these two elements. (I. 2.)

V.

Fact and *Theory* correspond to Sense on the one hand, and to Ideas on the other, so far as we are *conscious* of our Ideas: but all facts involve ideas *unconsciously*; and thus the distinction of Facts and Theories is not tenable, as that of Sense and Ideas is. (I. 2.)

VI.

Sensations and Ideas in our knowledge are like Matter and Form in bodies. Matter cannot exist without Form, nor Form without Matter: yet the two are altogether distinct and opposite. There is no possibility either of separating, or of confounding them. The same is the case with Sensations and Ideas. (I. 2.)

VII.

Ideas are not *trans*formed, but *in*formed Sensations; for without ideas, sensations have no form. (I. 2.)

VIII.

The Sensations are the *Objective*, the Ideas the *Subjective* part of every act of perception or knowledge. (I. 2.)

IX.

General Terms denote *Ideal Conceptions*, as a *circle*, an *orbit*, a *rose*. These are not *Images* of real things, as was held by the Realists, but Conceptions: yet they are conceptions, not bound together by mere *Name*, as the Nominalists held, but by an Idea. (I. 2.)

X.

It has been said by some, that all Conceptions are merely *states* or *feelings of the mind*, but this assertion only tends to confound what it is our business to distinguish. (I. 2.)

XI.

Observed Facts are connected so as to produce new truths, by superinducing upon them an Idea: and such truths are obtained *by Induction*. (I. 2.)

XII.

Truths once obtained by legitimate Induction are Facts: these Facts may be again connected, so as to produce higher truths: and thus we advance to *Successive Generalizations*. (I. 2.)

LXXIII. (*Doubtful.*)

Coexistent polarities are fundamentally identical. (v. 2.)

LXXIV.

The Idea of Chemical *Affinity*, as implied in Elementary Composition, involves peculiar conceptions. It is not properly expressed by assuming the qualities of bodies to *resemble* those of the elements, or to depend on the *figure* of the elements, or on their *attractions*. (vi. 1.)

LXXV.

Attractions take place between bodies, Affinities between the particles of a body. The former may be compared to the alliances of states, the latter to the ties of family. (vi. 2.)

LXXVI.

The governing principles of Chemical Affinity are, that it is *elective;* that it is *definite;* that it *determines the properties* of the compound; and that *analysis is possible.* (vi. 2.)

LXXVII.

We have an idea of *Substance:* and an axiom involved in this Idea is, that *the weight of a body is the sum of the weights of all its elements.* (vi. 3).

LXXVIII.

Hence Imponderable Fluids are not to be admitted as chemical elements. (vi. 4.)

LXXIX.

The Doctrine of Atoms is admissible as a mode of expressing and calculating laws of nature; but is not proved by any fact, chemical or physical, as a philosophical truth. (vi. 5.)

LXXX.

We have an Idea of *Symmetry;* and an axiom involved in this Idea is, that a symmetrical natural body, if there be a tendency to modify any member in any manner, there is a tendency to modify all the corresponding members in the same manner. (VII. 1.)

LXXXI.

All hypotheses respecting the manner in which the elements of inorganic bodies are arranged in space, must be constructed with regard to the general facts of crystallization. (VII. 3.)

LXXXII.

When we consider any object as *One,* we give unity to it by an act of thought. The condition which determines what this unity shall include, and what it shall exclude, is this;—that assertions concerning the one thing shall be possible. (VIII. 1.)

LXXXIII.

We collect individuals into *Kinds* by applying to them the Idea of Likeness. Kinds of things are not determined by definitions, but by this condition;—that general assertions concerning such kinds of things shall be possible. (VIII. 1.)

LXXXIV.

The *Names* of kinds of things are governed by their use; and that may be a right name in one use which is not so in another. A whale is not a *fish* in natural history, but it is a *fish* in commerce and law. (VIII. 1.)

LXXXV.

We take for granted that each kind of things has a special *character* which may be expressed by a Definition. The ground of our assumption is this;—that reasoning must be possible. (VIII. 1.)

LXXXVI.

The "Five Words," *Genus, Species, Difference, Property, Accident,* were used by the Aristotelians, in order to express the subordination of kinds, and to describe the nature of definitions and propositions. In modern times, these technical expressions have been more referred to by Natural Historians than by Metaphysicians. (VIII. 1.)

LXXXVII.

The construction of a Classificatory Science includes *Terminology,* the formation of a descriptive language; —*Diataxis,* the Plan of the System of Classification, called also the *Systematick;—Diagnosis,* the Scheme of the Characters by which the different Classes are known, called also the *Characteristick. Physiography* is the knowledge which the System is employed to convey. Diataxis includes *Nomenclature.* (VIII. 2.)

LXXXVIII.

Terminology must be conventional, precise, constant; copious in words, and minute in distinctions, according to the needs of the science. The student must understand the terms, *directly* according to the convention, not through the medium of explanation or comparison. (VIII. 2.)

LXXXIX.

The *Diataxis,* or Plan of the System, may aim at a Natural or an Artificial System. But no classes can be absolutely artificial, for if they were, no assertions could be made concerning them. (VIII. 2.)

APHORISMS CONCERNING SCIENCE.

I.

THE two processes by which Science is constructed are the *Explication of Conceptions* and the *Colligation of Facts*. (Book XI. Chap. 1.)

II.

The Explication of Conceptions, as requisite for the progress of science, has been effected by means of discussions and controversies among scientists; often by debates concerning definitions; these controversies have frequently led to the establishment of a Definition; but along with the Definition, a corresponding Proposition has always been expressed or implied. The essential requisite for the advance of science is the clearness of the Conception, not the establishment of a Definition. The construction of an exact Definition is often very difficult. The requisite conditions of clear Conceptions may often be expressed by Axioms as well as by Definitions. (XI. 2.)

III.

Conceptions, for purposes of science, must be *appropriate* as well as clear: that is, they must be modifications of that Fundamental Idea, by which the phenomena can really be interpreted. This maxim may warn us from errour, though it may not lead to discovery. Discovery depends upon the previous cultivation or natural clearness of the appropriate Idea, and therefore *no discovery is the work of accident*. (XI. 2.)

Facts are the materials of science, but all Facts involve Ideas. Since, in observing Facts, we cannot exclude Ideas, we must, for the purposes of science, take care that the Ideas are clear and rigorously applied. (XI. 3.)

V.

The last Aphorism leads to such Rules as the following:—That Facts, for the purposes of material science, must involve Conceptions of the Intellect only, and not Emotions:—That Facts must be observed with reference to our most exact conceptions, Number, Place, Figure, Motion:—That they must also be observed with reference to any other exact conceptions which the phenomena suggest, as Force, in mechanical phenomena, Concord, in musical. (XI. 3.)

VI.

The resolution of complex Facts into precise and measured partial Facts, we call the *Decomposition of Facts*. This process is requisite for the progress of science, but does not necessarily lead to progress. (XI. 3.)

VII.

Science begins with *common* observation of facts; but even at this stage, requires that the observations be precise. Hence the sciences which depend upon space and number were the earliest formed. After common observation, come Scientific *Observation* and *Experiment*. (XI. 4.)

VIII.

The Conceptions by which Facts are bound together, are suggested by the sagacity of discoverers. This sagacity cannot be taught. It commonly succeeds by guessing; and this success seems to consist in framing several *tentative hypotheses* and selecting the right one. But a supply of appropriate hypotheses cannot be constructed by rule, nor without inventive talent. (XI. 4.)

IX.

The truth of tentative hypotheses must be tested by their application to facts. The discoverer must be ready, carefully to try his hypotheses in this manner, and to reject them if they will not bear the test, in spite of indolence and vanity. (XI. 4.)

X.

The process of scientific discovery is cautious and rigorous, not by abstaining from hypotheses, but by rigorously comparing hypotheses with facts, and by resolutely rejecting all which the comparison does not confirm. (XI. 5.)

XI.

Hypotheses may be useful, though involving much that is superfluous, and even erroneous: for they may supply the true bond of connexion of the facts; and the superfluity and errour may afterwards be pared away. (XI. 5.)

XII.

It is a test of true theories not only to account for, but to predict phenomena. (XI. 5.)

XIII.

Induction is a term applied to describe the *process* of a true Colligation of Facts by means of an exact and appropriate Conception. *An Induction* is also employed to denote the *proposition* which results from this process. (XI. 5.)

XIV

The Consilience of Inductions takes place when an Induction, obtained from one class of facts, coincides with an Induction, obtained from another different class. This Consilience is a test of the truth of the Theory in which it occurs.   (XI. 5.)

XV.

An Induction is not the mere *sum* of the Facts which are colligated.   The Facts are not only brought together, but seen in a new point of view.   A new mental Element is *superinduced*; and a peculiar constitution and discipline of mind are requisite in order to make this Induction.   (XI. 5.)

XVI.

Although in Every Induction a new conception is superinduced upon the Facts; yet this once effectually done, the novelty of the conception is overlooked, and the conception is considered as a part of the fact. (XI. 5.)

XVII.

The *Logic of Induction* consists in stating the Facts and the Inference in such a manner, that the Evidence of the Inference is manifest; just as the Logic of Deduction consists in stating the Premises and the Conclusion in such a manner that the Evidence of the Conclusion is manifest.   (XI. 6.)

XVIII.

The Logic of Deduction is exhibited by means of a certain Formula; namely, a Syllogism; and every train of deductive reasoning, to be demonstrative, must be capable of resolution into a series of such Formulæ legitimately constructed.   In like manner, the Logic of Induction may be exhibited by means of certain *Formulæ*;

and every train of inductive inference, to be sound, must be capable of resolution into a scheme of such Formulæ, legitimately constructed.   (XI. 6.)

XIX.

The *inductive act of thought* by which several Facts are colligated into one Proposition, may be expressed by saying: *The several Facts are exactly expressed as one Fact, if, and only if, we adopt the Conceptions and the Assertion* of the Proposition.   (XI. 6.)

XX.

The One Fact, thus inductively obtained from several Facts, may be combined with other Facts, and colligated with them by a new act of Induction.   This process may be indefinitely repeated: and these successive processes are the *Steps* of Induction, or of *Generalization*, from the lowest to the highest.   (XI. 6.)

XXI.

The relation of the successive Steps of Induction may be exhibited by means of an *Inductive Table*, in which the several Facts are indicated, and tied together by a Bracket, and the Inductive Inference placed on the other side of the Bracket; and this arrangement repeated, so as to form a genealogical Table of each Induction, from the lowest to the highest.   (XI. 6.)

XXII.

The Logic of Induction is the *Criterion of Truth* inferred from Facts, as the Logic of Deduction is the Criterion of Truth deduced from necessary Principles. The Inductive Table enables us to apply such a Criterion; for we can determine whether each Induction is verified and justified by the Facts which its Bracket includes; and if each induction in particular be sound, the highest, which merely combines them all, must necessarily be sound also. (XI. 6.)

XXIII.

The distinction of *Fact* and *Theory* is only relative. Events and phenomena, considered as particulars which may be colligated by Induction, are *Facts;* considered as generalities already obtained by colligation of other Facts, they are *Theories.* The same event or phenomenon is a Fact or a Theory, according as it is considered as standing on one side or the other of the Inductive Bracket. (XI. 6.)

XXIV.

Inductive truths are of two kinds, *Laws of Phenomena*, and *Theories of Causes.* It is necessary to begin in every science with the Laws of Phenomena; but it is impossible that we should be satisfied to stop short of a Theory of Causes. In Physical Astronomy, Physical Optics, Geology, and other sciences, we have instances showing that we can make a great advance in inquiries after true Theories of Causes. (XI. 7.)

XXV.

Art and Science differ. The object of Science is Knowledge; the objects of Art, are Works. In Art, truth is a means to an end; in Science, it is the only end. Hence the Practical Arts are not to be classed among the Sciences. (XI. 8.)

XXVI.

Practical Knowledge, such as Art implies, is not Knowledge such as Science includes. Brute animals have a practical knowledge of relations of space and force; but they have no knowledge of Geometry or Mechanics. (XI. 8.)

ASTRONOMY AND GENERAL PHYSICS

CONSIDERED WITH REFERENCE TO

NATURAL THEOLOGY

BY THE

REV. WILLIAM WHEWELL M.A.

FELLOW AND TUTOR OF TRINITY COLLEGE

CAMBRIDGE

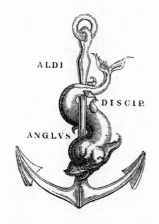

ALDI

DISCIP.

ANGLVS

LONDON
WILLIAM PICKERING
1834

ASTRONOMY

GENERAL PHYSICS.

―――――

INTRODUCTION.

CHAPTER I.

Object of the Present Treatise.

THE examination of the material world brings
before us a number of things and relations of
things which suggest to most minds the belief of
a creating and presiding Intelligence. And this
impression, which arises with the most vague and
superficial consideration of the objects by which
we are surrounded, is, we conceive, confirmed
and expanded by a more exact and profound
study of external nature. Many works have been
written at different times with the view of showing how our knowledge of the elements and their
operation, of plants and animals and their construction, may serve to nourish and unfold our

W.   B

idea of a Creator and Governor of the world. But though this is the case, a new work on the same subject may still have its use. Our views of the Creator and Governor of the world, as collected from or combined with our views of the world itself, undergo modifications, as we are led by new discoveries, new generalizations, to regard nature in a new light. The conceptions concerning the Deity, his mode of effecting his purposes, the scheme of his government, which are suggested by one stage of our knowledge of natural objects and operations, may become manifestly imperfect or incongruous, if adhered to and applied at a later period, when our acquaintance with the immediate causes of natural events has been greatly extended. On this account it may be interesting, after such an advance, to show how the views of the creation, preservation, and government of the universe, which natural science opens to us, harmonize with our belief in a Creator, Governor, and Preserver of the world. To do this with respect to certain departments of Natural Philosophy is the object of the following pages; and the author will deem himself fortunate, if he succeeds in removing any of the difficulties and obscurities which prevail in men's minds, from the want of a clear mutual understanding between the religious and the scientific speculator. It is needless here to remark the necessarily imperfect and scanty character of Natural Religion;

for most persons will allow that, however imperfect may be the knowledge of a Supreme Intelligence which we gather from the contemplation of the natural world, it is still of most essential use and value. And our purpose on this occasion is, not to show that Natural Theology is a perfect and satisfactory scheme, but to bring up our Natural Theology to the point of view in which it may be contemplated by the aid of our Natural Philosophy.

Now the peculiar point of view which at present belongs to Natural Philosophy, and especially to the departments of it which have been most successfully cultivated, is, that nature, so far as it is an object of scientific research, is a collection of facts governed by *laws:* our knowledge of nature is our knowledge of laws ; of laws of operation and connexion, of laws of succession and co-existence, among the various elements and appearances around us. And it must therefore here be our aim to show how this view of the universe falls in with our conception of the Divine Author, by whom we hold the universe to be made and governed.

Nature acts by general laws; that is, the occurrences of the world in which we find ourselves, result from causes which operate according to fixed and constant rules. The succession of days, and seasons, and years, is produced by the motions of the earth ; and these again are governed by the attraction of the sun, a force which

acts with undeviating steadiness and regularity.
The changes of winds and skies, seemingly so
capricious and casual, are produced by the operation of the sun's heat upon air and moisture,
land and sea ; and though in this case we cannot
trace the particular events to their general causes,
as we can trace the motions of the sun and moon,
no philosophical mind will doubt the generality
and fixity of the rules by which these causes
act. The variety of the effects takes place, because the circumstances in different cases vary ;
and not because the action of material causes
leaves anything to chance in the result. And
again, though the vital movements which go on
in the frame of vegetables and animals depend on
agencies still less known, and probably still more
complex, than those which rule the weather, each
of the powers on which such movements depend
has its peculiar laws of action, and these are as
universal and as invariable as the law by which
a stone falls to the earth when not supported.

The world then is governed by general laws ;
and in order to collect from the world itself a
judgment concerning the nature and character
of its government, we must consider the import
and tendency of such laws, so far as they come
under our knowledge. If there be, in the administration of the universe, intelligence and
benevolence, superintendence and foresight,
grounds for love and hope, such qualities may

be expected to appear in the constitution and combination of those fundamental regulations by which the course of nature is brought about, and made to be what it is.

If a man were, by some extraordinary event, to find himself in a remote and unknown country, so entirely strange to him that he did not know whether there existed in it any law or government at all; he might in no long time ascertain whether the inhabitants were controlled by any superintending authority; and with a little attention he might determine also whether such authority were exercised with a prudent care for the happiness and well-being of its subjects, or without any regard and fitness to such ends; whether the country were governed by laws at all, and whether the laws were good. And according to the laws which he thus found prevailing, he would judge of the sagacity, and the purposes of the legislative power.

By observing the laws of the material universe and their operation, we may hope, in a somewhat similar manner, to be able to direct our judgment concerning the government of the universe: concerning the mode in which the elements are regulated and controlled, their effects combined and balanced. And the general tendency of the results thus produced may discover to us something of the character of the power which has legislated for the material world.

We are not to push too far the analogy thus suggested. There is undoubtedly a wide difference between the circumstances of man legislating for man, and God legislating for matter. Still we shall, it will appear, find abundant reason to admire the wisdom and the goodness which have established *the Laws of Nature,* however rigorously we may scrutinize the import of this expression.

BOOK III.

THE contemplation of the material universe exhibits God to us as the author of the laws of material nature; bringing before us a wonderful spectacle, in the simplicity, the comprehensiveness, the mutual adaptation of these laws, and in the vast variety of harmonious and beneficial effects produced by their mutual bearing and combined operation. But it is the consideration of the moral world, of the results of our powers of thought and action, which leads us to regard the Deity in that light in which our relation to him becomes a matter of the highest interest and importance. We perceive that man is capable of referring his actions to principles of right and wrong; that both his faculties and his virtues may be unfolded and advanced by the discipline which arises from the circumstances of human society; that good men can be discriminated from the bad, only by a course of trial, by struggles with difficulty and temptation; that the best men feel deeply the need of relying, in such conflicts, on the thought of a superintending Spiritual Power; that our views of justice, our capacity for intellectual and moral advancement, and a crowd of hopes and anticipations which rise in our bosoms unsought, and cling there with inexhaustible tenacity, will not allow us to acquiesce in the belief that this life is the end of our existence. We are thus led to see that our relation to the Superintender of our moral being, to the Depositary of the supreme law of just and right, is a relation of incalculable consequence. We find that we cannot be permitted to be merely contemplators and speculators with regard to the Governor of the moral world; we must obey His will; we must turn our affections to Him; we must advance in His favour; or we offend against the nature of our position in the scheme of which He is the author and sustainer.

It is far from our purpose to represent natural religion as of itself sufficient for our support and guidance; or to underrate the manner in which our views of the Lord of the universe have been, much more, perhaps, than we are sometimes aware, illustrated and confirmed by lights derived from revelation. We do not here speak of the manner in which men have come to believe in God, as the Governor of the moral world; but of the fact, that by the aid of one or both of these two guides, Reason or Revelation, reflecting persons in every age have been led to such a belief. And we conceive it may be useful to point out some connexion between such a belief of a just and holy Governor, and the conviction, which we have already endeavoured to impress upon the reader, of a wise and benevolent Creator of the physical world. This we shall endeavour to do in the present book.

At the same time that men have thus learnt to look upon God as their Governor and Judge, the source of their support and reward, they have also been led, not only to ascribe to him power and skill, knowledge and goodness, but also to attribute to him these qualities in a mode and degree excluding all limit:—to consider him as almighty, allwise, of infinite knowledge and in-exhaustible goodness; every where present and active, but incomprehensible by our minds, both in the manner of his agency, and the degree of his perfections. And this impression concerning the Deity appears to be that which the mind receives from all objects of contemplation and all modes of advance towards truth. To this con-ception it leaps with alacrity and joy, and in this it acquiesces with tranquil satisfaction and grow-ing confidence; while any other view of the nature of the Divine Power which formed and sustained the world, is incoherent and untenable, exposed to insurmountable objections and in-tolerable incongruities. We shall endeavour to show that the modes of employment of the thoughts to which the well conducted study of nature gives rise, do tend, in all their forms, to produce or strengthen this impression on the mind ; and that such an impression, and no other, is consistent with the widest views and most comprehensive aspects of nature and of philosophy, which our Natural Philosophy opens to us. This will be the purpose of the latter part of the present book. In the first place we shall proceed with the object first mentioned, the connexion which may be perceived between the evidences of creative power, and of moral government, in the world.

CHAPTER I.

The Creator of the Physical World is the Governor of the Moral World.

WITH our views of the moral government of the world and the religious interests of man, the study of material nature is not and cannot be directly and closely connected. But it may be of some service to trace in these two lines of reasoning, seemingly so remote, a manifest convergence to the same point, a demonstrable unity of result. It may be useful to show that we are thus led, not to two rulers of the universe, but to one God ;—to make it appear that the Creator and Preserver of the world is also the Governor and Judge of men ; that the Author of the Laws of Nature is also the Author of the

Law of Duty ;—that He who regulates corporeal things by properties of attraction and affinity and assimilating power, is the same being who regulates the actions and conditions of men, by the influence of the feeling of responsibility, the perception of right and wrong, the hope of happiness, the love of good.

The conviction that the Divine attributes which we are taught by the study of the material world, and those which we learn from the contemplation of man as a responsible agent, belong to the same Divine Being, will be forced upon us, if we consider the manner in which all the parts of the universe, the corporeal and intellectual, the animal and moral, are connected with each other. In each of these provinces of creation we trace refined adaptations and arrangements which lead us to the Creator and Director of so skilful a system ; but these provinces are so intermixed, these different trains of contrivance so interwoven, that we cannot, in our thoughts, separate the author of one part from the author of another. The Creator of the Heavens and of the Earth, of the inorganic and of the organic world, of animals and of man, of the affections and the conscience, appears inevitably to be one and the same God.

We will pursue this reflection a little more into detail.

1. The *Atmosphere* is a mere mass of fluid

floating on the surface of the ball of the earth; it is one of the inert and inorganic portions of the universe, and must be conceived to have been formed by the same Power which formed the solid mass of the earth and all other parts of the solar system. But how far is the atmosphere from being inert in its effects on organic beings, and unconnected with the world of life! By what wonderful adaptations of its mechanical and chemical properties, and of the vital powers of plants, to each other, are the developement and well-being of plants and animals secured! The creator of the atmosphere must have been also the creator of plants and animals: we cannot for an instant believe the contrary. But the atmosphere is not only subservient to the life of animals, and of man among the rest; it is also the vehicle of voice; it answers the purpose of intercourse; and, in the case of man, of rational intercourse. We have seen how remarkably the air is fitted for this office; the construction of the organs of articulation, by which they are enabled to perform their part of the work, is, as is well known, a most exquisite system of contrivances. But though living in an atmosphere capable of transmitting articulate sound, and though provided with organs fitted to articulate, man would never attain to the use of language, if he were not also endowed with another set of faculties. The powers of abstraction and generalization, memory and reason, the tendencies which occasion the inflexions and combinations of words, are all necessary to the formation and use of language. Are not these parts of the same scheme of which the bodily faculties by which we are able to speak are another part? Has man his mental powers independently of the creator of his bodily frame? To what purpose then, or by what cause was the curious and complex machinery of the tongue, the glottis, the larynx produced? These are useful for speech, and full of contrivances which suggest such a use as the end for which those organs were constructed. But speech appears to have been no less contemplated in the intellectual structure of man. The processes of which we have spoken, generalization, abstraction, reasoning, have a close dependence on the use of speech. These faculties are presupposed in the formation of language, but they are developed and perfected by the use of language. The mind of man then, with all its intellectual endowments, is the work of the same artist by whose hands his bodily frame was fashioned; as his bodily faculties again are evidently constructed by the maker of those elements on which their action depends. The creator of the atmosphere and of the material universe is the creator of the human mind, and the author of those wonderful powers of thinking, judging, inferring, discovering, by which we are able to

w.   s

reason concerning the world in which we are placed ; and which aid us in lifting our thoughts to the source of our being himself.

2. *Light*, or the means by which light is propagated, is another of the inorganic elements which forms a portion of the mere material world. The luminiferous ether, if we adopt that theory, or the fluid light of the theory of emission, must indubitably pervade the remotest regions of the universe, and must be supposed to exist, as soon as we suppose the material parts of the universe to be in existence. The origin of light then must be at least as far removed from us as the origin of the solar system. Yet how closely connected are the properties of light with the structure of our own bodies! The mechanism of the organs of vision and the mechanism of light are, as we have seen, most curiously adapted to each other. We must suppose, then, that the same power and skill produced one and the other of these two sets of contrivances, which so remarkably *fit into* each other. The creator of light is the author of our visual powers. But how small a portion does mere visual perception constitute of the advantages which we derive from vision! We possess ulterior faculties and capacities by which sight becomes a source of happiness and good to man. The sense of beauty, the love of art, the pleasure arising from the contemplation of nature, are all dependent on

the eye ; and we can hardly doubt that these faculties were bestowed on man to further the best interests of his being. The sense of beauty both animates and refines his domestic tendencies ; the love of art is a powerful instrument for raising him above the mere cravings and satisfactions of his animal nature ; the expansion of mind which rises in us at the sight of the starry sky, the cloud-capt mountain, the boundless ocean, seems intended to direct our thoughts by an impressive though indefinite feeling, to the Infinite Author of All. But if these faculties be thus part of the scheme of man's inner being, given him by a good and wise creator, can we suppose that this creator was any other than the creator also of those visual organs, without which the faculties could have no operation and no existence ? As clearly as light and the eye are the work of the same author, so clearly also do our capacities for the most exalted visual pleasures, and the feelings flowing from them, proceed from the same Divine Hand, by which the mechanism of light was constructed.

3. The creator of the earth must be conceived to be the author also of all those qualities in the soil, chemical and whatever else, by which it supports vegetable life, under all the modifications of natural and artificial condition. Among the attributes which the earth thus possesses, there are some which seem to have an especial

reference to man in a state of society. Such are
the power of the earth to increase its produce
under the influence of cultivation, and the necessary existence of property in land, in order
that this cultivation may be advantageously
applied; the rise, under such circumstances, of
a *surplus* produce, of a quantity of subsistence
exceeding the wants of the cultivators alone;
and the consequent possibility of inequalities of
rank and of all the arrangements of civil society.
These are all parts of the constitution of the
earth. But these would all remain mere idle
possibilities, if the nature of man had not a corresponding direction. If man had not a social
and economical tendency, a disposition to con-
gregate and cooperate, to distribute possessions
and offices among the members of the community,
to make and obey and enforce laws, the earth
would in vain be ready to respond to the care of
the husbandman. Must we not then suppose that
this attribute of the earth was bestowed upon
it by Him who gave to man those corresponding
attributes, through which the apparent niggard-
liness of the soil is the source of general comfort
and security, of polity and law? Must we not
suppose that He who created the soil also in-
spired man with those social desires and feelings
which produce cities and states, laws and institutions, arts and civilization; and that thus the
apparently inert mass of earth is a part of the

same scheme as those faculties and powers with which man's moral and intellectual progress is most connected?

4. Again:—It will hardly be questioned that the author of the material elements is also the author of the structure of animals, which is adapted to and provided for by the constitution of the elements in such innumerable ways. But the author of the bodily structure of animals must also be the author of their instincts, for without these the structure would not answer its purpose. And these instincts frequently assume the character of affections in a most remarkable manner. The love of offspring, of home, of companions, are often displayed by animals, in a way that strikes the most indifferent observer; and yet these affections will hardly be denied to be a part of the same scheme as the instincts by which the same animals seek food and the gratifications of sense. Who can doubt that the anxious and devoted affection of the mother-bird for her young after they are hatched, is a part of the same system of Providence as the instinct by which she is impelled to sit upon her eggs? and this, of the same by which her eggs are so organized that incubation leads to the birth of the young animal? Nor, again, can we imagine that while the structure and affections of animals belong to one system of things, the affections of man, in many respects so similar to those of animals, and

connected with the bodily frame in a manner so closely analogous, can belong to a different scheme. Who, that reads the touching instances of maternal affection, related so often of the women ,of all nations, and of the females of all animals, can doubt that the principle of action is the same in the two cases, though enlightened in one of them by the rational faculty? And who can place in separate provinces the supporting and protecting love of the father and of the mother? or consider as entirely distinct from these, and belonging to another part of our nature, the other kinds of family affection? or disjoin man's love of his home, his clan, his tribe, his country, from the affection which he bears to his family? The love of offspring, home, friends, in man, is then part of the same system of contrivances of which bodily organization is another part. And thus the author of our corporeal frame is also the author of our capacity of kindness and resentment, of our love and of our wish to be loved, of all the emotions which bind us to individuals, to our families, and to our kind.

It is not necessary here to follow out and classify these emotions and affections; or to examine how they are combined and connected with our other motives of action, mutually giving and receiving strength and direction. The desire of esteem, of power, of knowledge, of society, the love of kindred, of friends, of our country, are manifestly among the main forces by which man is urged to act and to abstain. And as these parts of the constitution of man are clearly intended, as we conceive, to impel him in his appointed path; so we conceive that they are no less clearly the work of the same great Artificer who created the heart, the eye, the hand, the tongue, and that elemental world in which, by means of these instruments, man pursues the objects of his appetites, desires, and affections.

5. But if the Creator of the world be also the author of our intellectual powers, of our feeling for the beautiful and the sublime, of our social tendencies, and of our natural desires and affections, we shall find it impossible not to ascribe also to Him the higher directive attributes of our nature, the conscience and the religious feeling, the reference of our actions to the rule of duty and to the will of God.

It would not suit the plan of the present treatise to enter into any detailed analysis of the connexion of these various portions of our moral constitution. But we may observe that the existence and universality of the conception of duty and right cannot be doubted, however men may differ as to its original or derivative nature. All men are perpetually led to form judgments concerning actions, and emotions which lead to action, as right or wrong; as what they *ought* or *ought not* to do or feel. There is a faculty which approves and disapproves, acquits or condemns the workings of our other faculties. Now, what shall we say of such a judiciary principle, thus introduced among our motives to action? Shall we conceive that while the other springs of action are balanced against each other by our Creator, this, the most pervading and universal regulator, was no part of the original scheme? That— while the love of animal pleasures, of power, of fame, the regard for friends, the pleasure of bestowing pleasure, were infused into man as influences by which his course of life was to be carried on, and his capacities and powers developed and exercised;—this reverence for a moral law, this acknowledgment of the obligation of duty,—a feeling which is everywhere found, and which may become a powerful, a predominating motive of action,—was given for no purpose, and belongs not to the design? Such an opinion would be much as if we should acknowledge the skill and contrivance manifested in the other parts of a ship, but should refuse to recognize the rudder as exhibiting any evidence of a purpose. Without the reverence which the opinion of right inspires, and the scourge of general disapprobation inflicted on that which is accounted wicked, society could scarcely go on; and certainly the feelings and thoughts and characters of men could not be what they are. Those impulses of nature which involve no acknowledgment of responsibility, and the play and struggle of interfering wishes, might preserve the species in some shape of existence, as we see in the case of brutes. But a person must be strangely constituted, who, living amid the respect for law, the admiration for what is good, the order and virtues and graces of civilized nations, (all which have their origin in some degree in the feeling of responsibility) can maintain that all these are casual and extraneous circumstances, no way contemplated in the formation of man ; and that a condition in which there should be no obligation in law, no merit in self-restraint, no beauty in virtue, is equally suited to the powers and the nature of man, and was equally contemplated when those powers were given him.

If this supposition be too extravagant to be admitted, as it appears to be, it remains then that man, intended, as we have already seen from his structure and properties, to be a discoursing, social being, acting under the influence of affections, desires, and purposes, was also intended to act under the influence of a sense of duty ; and that the acknowledgment of the obligation of a moral law is as much part of his nature, as hunger or thirst, maternal love or the desire of power; that, therefore, in conceiving man as the work of a Creator, we must imagine his powers and character given him with an intention on the Creator's part that this sense of duty should occupy its place in his constitution as an active and thinking being : and that this directive and judiciary principle is a part of the work of the same Author who made the elements to minister to the material functions, and the arrangements of the world to occupy the individual and social affections of his living creatures.

This principle of conscience, it may further be observed, does not stand upon the same level as the other impulses of our constitution by which we are prompted or restrained. By its very nature and essence, it possesses a supremacy over all others. " Your obligation to obey this law is its being the law of your nature. That your conscience approves of and attests such a course of action is itself alone an obligation. Conscience does not only offer itself to show us the way we should walk in, but it likewise carries its own authority with it, that it is our natural guide : the guide assigned us by the author of our nature."* That we ought to do an action, is of itself a sufficient and ultimate answer to the questions, *why* we should do it ?—how we are *obliged* to do it ? The conviction of duty implies the soundest reason, the strongest obligation, of which our nature is susceptible.

We appear then to be using only language which is well capable of being justified, when we speak of this irresistible esteem for what is

* Butler, Serm. 3.

right, this conviction of a rule of action extending beyond the gratification of our irreflective impulses, as an impress stamped upon the human mind by the Deity himself; a trace of His nature; an indication of His will; an announcement of His purpose; a promise of His favour; and though this faculty may need to be confirmed and unfolded, instructed and assisted by other aids, it still seems to contain in itself a sufficient intimation that the highest objects of man's existence are to be attained, by means of a direct and intimate reference of his thoughts and actions to the Divine Author of his being.

Such then is the Deity to which the researches of Natural Theology point; and so far is the train of reflections in which we have engaged, from being merely speculative and barren. With the material world we cannot stop. If a superior Intelligence *have* ordered and adjusted the succession of seasons and the structure of the plants of the field, we must allow far more than this at first sight would seem to imply. We must admit still greater powers, still higher wisdom for the creation of the beasts of the forest with their faculties; and higher wisdom still and more transcendent attributes, for the creation of man. And when we reach this point, we find that it is not knowledge only, not power only, not foresight and beneficence alone, which we must attribute to the Maker of the World; but that we must consider him as the Author, in us, of a reverence for moral purity and rectitude; and, if the author of such emotions in us, how can we conceive of Him otherwise, than that these qualities are parts of his nature; and that he is not only wise and great, and good, incomparably beyond our highest conceptions, but also conformed in his purposes to the rule which he thus impresses upon us, that is, Holy in the highest degree which we can image to ourselves as possible.

Chapter II.

On the Vastness of the Universe.

1. The aspect of the world, even without any of the peculiar lights which science throws upon it, is fitted to give us an idea of the greatness of the power by which it is directed and governed, far exceeding any notions of power and greatness which are suggested by any other contemplation. The number of human beings who surround us— the various conditions requisite for their life, nutrition, well-being, all fulfilled;—the way in which these conditions are modified, as we pass in thought to other countries, by climate, temperament, habit;—the vast amount of the human population of the globe thus made up;—yet man himself but one among almost endless tribes of

animals ;—the forest, the field, the desert, the air, the ocean, all teeming with creatures whose bodily wants are as carefully provided for as his ;—the sun, the clouds, the winds, all attending, as it were, on these organized beings ;—a host of beneficent energies, unwearied by time and succession, pervading every corner of the earth ;—this spectacle cannot but give the contemplator a lofty and magnificent conception of the Author of so vast a work, of the Ruler of so wide and rich an empire, of the Provider for so many and varied wants, the Director and Adjuster of such complex and jarring interests.

But when we take a more exact view of this spectacle, and aid our vision by the discoveries which have been made of the structure and extent of the universe, the impression is incalculably increased.

The number and variety of animals, the exquisite skill displayed in their structure, the comprehensive and profound relations by which they are connected, far exceed any thing which we could have beforehand imagined. But the view of the universe expands also on another side. The earth, the globular body thus covered with life, is not the only globe in the universe. There are, circling about our own sun, six others, so far as we can judge, perfectly analogous in their nature : besides our moon and other bodies analogous to it. No one can resist the tempta-

tion to conjecture, that these globes, some of
them much larger than our own, are not dead
and barren ;—that they are, like ours, occupied
with organization, life, intelligence. To conjecture is all that we can do, yet even by the
perception of such a possibility, our view of the
domain of nature is enlarged and elevated. The
outermost of the planetary globes of which we
have spoken is so far from the sun, that the
central luminary must appear to the inhabitants
of that planet, if any there are, no larger than
Venus does to us ; and the length of their year
will be 82 of ours.

But astronomy carries us still onwards. It
teaches us that, with the exception of the planets
already mentioned, the stars which we see have
no immediate relation to our system. The obvious supposition is that they are of the nature
and order of our sun : the minuteness of their
apparent magnitude agrees, on this supposition,
with the enormous and almost inconceivable
distance which, from all the measurements of
astronomers, we are led to attribute to them. If
then these are suns, they may, like our sun, have
planets revolving round them; and these may,
like our planet, be the seats of vegetable and
animal and rational life :—we may thus have in
the universe worlds, no one knows how many, no
one can guess how varied ;—but however many,
however varied, they are still but so many pro-

vinces in the same empire, subject to common rules, governed by a common power.

But the stars which we see with the naked eye are but a very small portion of those which the telescope unveils to us. The most imperfect telescope will discover some that are invisible without it; the very best instrument perhaps does not show us the most remote. The number of stars which crowd some parts of the heavens is truly marvellous: Dr. Herschel calculated that a portion of the milky way, about 10 degrees long and 2½ broad, contained 258,000. In a sky so occupied the moon would eclipse 2000 of such stars at once.

We learn too from the telescope that even in this province the variety of nature is not exhausted. Not only do the stars differ in colour and appearance, but some of them grow periodically fainter and brighter, as if they were dark on one side, and revolved on their axes. In other cases two stars appear close to each other, and in some of these cases it has been clearly established, that the two have a motion of revolution about each other; thus exhibiting an arrangement new to the astronomer, and giving rise, possibly, to new conditions of worlds. In other instances again, the telescope shows, not luminous points, but extended masses of dilute light, like bright clouds, hence called *nebulæ*. Some have supposed (as we have noticed in the last book) that such nebulæ by further condensation might become suns; but for such opinions we have nothing but conjecture. Some stars again have undergone permanent changes; or have absolutely disappeared, as the celebrated star of 1572, in the constellation Cassiopea.

If we take the whole range of created objects in our own system, from the sun down to the smallest animalcule, and suppose such a system, or something in some way analogous to it, to be repeated for each of the millions of stars which the telescope reveals to us, we obtain a representation of the material universe; at least a representation which to many persons appears the most probable one. And if we contemplate this aggregate of systems as the work of a Creator, which in our own system we have found ourselves so irresistibly led to do, we obtain a sort of estimate of the extent through which his creative energy may be traced, by taking the widest view of the universe which our faculties have attained.

If we consider further the endless and admirable contrivances and adaptations which philosophers and observers have discovered in every portion of our own system; every new step of our knowledge showing us something new in this respect; and if we combine this consideration with the thought how small a portion of the universe our knowledge includes, we shall, without being able at all to discern the extent of the skill and wisdom displayed in the creation, see something of the character of the design, and of the copiousness and ampleness of the means which the scheme of the world exhibits. And when we see that the tendency of all the arrangements which we can comprehend is to support the existence, to develope the faculties, to promote the well-being of these countless species of creatures ; we shall have some impression of the beneficence and love of the Creator, as manifested in the physical government of his creation.

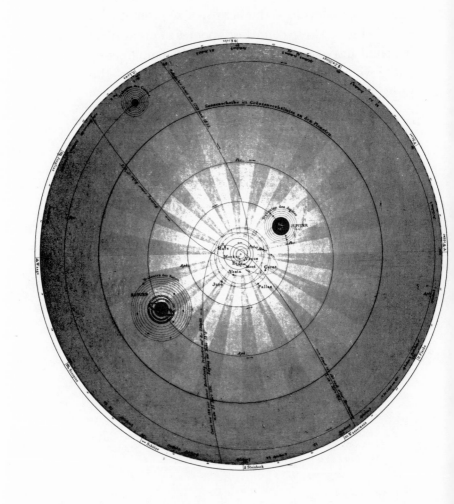

CHAPTER V.

On Inductive Habits; or, on the Impression produced on Men's Minds by discovering Laws of Nature.

THE object of physical science is to discover such laws and properties as those of which we have spoken in the last chapter. In this task, undoubtedly a progress has been made on which we may well look with pleasure and admiration; yet we cannot hesitate to confess that the extent of our knowledge on such subjects bears no proportion to that of our ignorance. Of the great and comprehensive laws which rule over the widest provinces of natural phenomena, few have yet been disclosed to us. And the names of the philosophers, whose high office it has been to detect such laws, are even yet far from numerous. In looking back at the path by which science has advanced to its present position, we see the names of the great discoverers shine out like luminaries, few and scattered along the line: by far the largest portion of the space is occupied by those whose comparatively humble office it was to verify, to develope, to apply the general truths which the discoverers brought to light.

It will readily be conceived that it is no easy matter, if it be possible, to analyse the process of thought by which laws of nature have thus been discovered; a process which, as we have said, has been in so few instances successfully performed. We shall not here make any attempt at such an analysis. But without this, we conceive it may be shown that the constitution and employment of the mind on which such discoveries depend, are friendly to that belief in a wise and good Creator and Governor of the world, which it has been our object to illustrate and confirm. And if it should appear that those who see further than their fellows into the bearings and dependencies of the material things and elements by which they are surrounded, have also been, in almost every case, earnest and forward in acknowledging the relation of all things to a supreme intelligence and will; we shall be fortified in our persuasion that the true scientific perception of the general constitution of the universe, and of the mode in which events are produced and connected, is fitted to lead us to the conception and belief of God.

Let us consider for a moment what takes place in the mind of a student of nature when he

attains to the perception of a law previously unknown, connecting the appearances which he
has studied. A mass of facts which before
seemed incoherent and unmeaning, assume, on a
sudden, the aspect of connexion and intelligible
order. Thus, when Kepler discovered the law
which connects the periodic times with the
diameters of the planetary orbits; or, when
Newton showed how this and all other known
mathematical properties of the solar system were
included in the law of universal gravitation according to the inverse square of the distance;
particular circumstances which, before, were
merely matter of independent record, became,
from that time, indissolubly conjoined by the laws
so discovered. The separate occurrences and
facts, which might hitherto have seemed casual
and without reason, were now seen to be all exemplifications of the same truth. The change is
like that which takes place when we attempt to
read a sentence written in difficult or imperfect
characters. For a time the separate parts appear to be disjointed and arbitrary marks; the
suggestions of possible meanings, which succeed
each other in the mind, fail, as fast as they are
tried, in combining or accounting for these symbols: but at last the true supposition occurs;
some words are found to coincide with the meaning thus assumed;. the whole line of letters appear to take definite shapes and to leap into

w.   x

their proper places ; and the truth of the happy conjecture seems to flash upon us from every part of the inscription.

The discovery of laws of nature, truly and satisfactorily connecting and explaining phenomena, of which, before, the connexion and causes had been unknown, displays much of a similar process, of obscurity succeeded by evidence, of effort and perplexity followed by conviction and repose. The innumerable conjectures and failures, the glimpses of light perpetually opening and as often clouded over, by which Kepler was tantalized, the unwearied perseverance and inexhaustible ingenuity which he exercised, while seeking for the laws which he finally discovered, are, thanks to his communicative disposition, curiously exhibited in his works, and have been narrated by his biographers ; and such efforts and alternations, modified by character and circumstances, must generally precede the detection of any of the wider laws and dependencies by which the events of the universe are regulated. We may readily conceive the satisfaction and delight with which, after this perplexity and struggle, the discoverer finds himself in light and tranquillity ; able to look at the province of nature which has been the subject of his study, and to read there an intelligible connexion, a sufficing reason, which no one before him had understood or apprehended.

This step so much resembles the mode in which one intelligent being understands and apprehends the conceptions of another, that we cannot be surprised if those persons in whose minds such a process has taken place, have been most ready to acknowledge the existence and operation of a superintending intelligence, whose ordinances it was their employment to study. When they had just read a sentence of the table of the laws of the universe, they could not doubt whether it had had a legislator. When they had decyphered there a comprehensive and substantial truth, they could not believe that the letters had been thrown together by chance. They could not but readily acknowledge that what their faculties had enabled them to read, must have been written by some higher and profounder mind. And accordingly, we conceive it will be found, on examining the works of those to whom we owe our knowledge of the laws of nature, and especially of the wider and more comprehensive laws, that such persons have been strongly and habitually impressed with the persuasion of a Divine Purpose and Power which had regulated the events which they had attended to, and ordained the laws which they had detected.

To those who have pursued science without reaching the rank of discoverers ;—who have possessed a derivative knowledge of the laws of nature which others had disclosed, and have employed themselves in tracing the consequences of such laws, and systematising the body of truth thus produced, the above description does not apply ; and we have not therefore in these cases the same ground for anticipating the same frame of mind. If among men of science of this class, the persuasion of a supreme Intelligence has at some periods been less vivid and less universal, than in that higher class of which we have before spoken, the fact, so far as it has existed, may perhaps be in some degree accounted for. But whether the view which we have to give of the mental peculiarities of men whose science is of this derivative kind be well founded, and whether the account we have above offered of that which takes place in the minds of original discoverers of laws in scientific researches be true, or not, it will probably be considered a matter of some interest to point out historically that in fact, such discoverers have been peculiarly in the habit of considering the world as the work of God. This we shall now proceed to do.

As we have already said, the names of *great* discoverers are not very numerous. The sciences which we may look upon as having reached or at least approached their complete and finished form, are Mechanics, Hydrostatics, and Physical Astronomy. Galileo is the father of modern Mechanics ; Copernicus, Kepler, and Newton are the great names which mark the progress of Astronomy. Hydrostatics shared in a great measure the fortunes of the related science of Mechanics; Boyle and Pascal were the persons mainly active in developing its more peculiar principles. The other branches of knowledge which belong to natural philosophy, as Chemistry and Meteorology, are as yet imperfect, and perhaps infant sciences; and it would be rash to presume to select, in them, names of equal preeminence with those above mentioned: but it may not be difficult to show, with sufficient evidence, that the effect of science upon the authors of science is, in these subjects as in the former ones, far other than to alienate their minds from religious trains of thought, and a habit of considering the world as the work of God.

We shall not dwell much on the first of the above mentioned great names, Galileo; for his scientific merit consisted rather in adopting the sound philosophy of others, as in the case of the Copernican system, and in combating prevalent errors, as in the case of the Aristotelian doctrines concerning motion, than in any marked and prominent discovery of new principles. Moreover the mechanical laws which he had a share in bringing to light, depending as they did, rather on detached experiments and transient facts, than on observation of the general course of the universe, could not so clearly suggest any reflexion on the government of the world at that period, as they did afterwards when Newton showed their bearing on the cosmical system. Yet Galileo, as a man of philosophical and inventive mind, who produced a great effect on the progress of physical knowledge, is a person whose opinions must naturally interest us, engaged in our present course of reasoning. There is in his writings little which bears upon religious views, as there is in the nature of his works little to lead him to such subjects. Yet strong expressions of piety are not wanting, both in his letters, and in his published treatises. The persecution which he underwent, on account of his writings in favour of the Copernican system, was grounded, not on his opposition to the general truths of natural religion, which is our main concern at present, nor even on any supposed rejection of any articles of Christian faith, but on the alleged discrepancy between his adopted astronomical views and the declarations of scripture. Some of his remarks may interest the reader.

In his third dialogue on the Copernican system he has occasion to speak of the opinion which holds all parts of the world to be framed for man's use alone : and to this he says, " I would that we should not so shorten the arm of God in the government of human affairs ; but that we should rest in this, that we are certain that God and nature are so occupied in the government of human affairs, that they could not more attend to us if they were charged with the care of the human race alone." In the same spirit, when some objected to the asserted smallness of the Medicean stars, or satellites of Jupiter, and urged this as a reason why they were unworthy the regard of philosophers, he replied that they are the works of God's power, the objects of His care, and therefore may well be considered as sublime subjects for man's study.

In the Dialogues on Mechanics, there occur those observations concerning the use of the air-bladder in fishes, and concerning the adaptation of the size of animals to the strength of the materials of which they are framed, which have often since been adopted by writers on the wisdom of Providence. The last of the dialogues on the system of the world is closed by a religious reflexion, put in the mouth of the interlocutor who usually expresses Galileo's own opinions. " While it is permitted us to speculate concerning the constitution of the world, we are also taught (perhaps in order that the activity of the human mind may not pause or languish) that our powers do not enable us to comprehend the works of His hands. May success therefore attend this intellectual exercise, thus permitted and appointed for us; by which we recognise and admire the greatness of God the more, in proportion as we find ourselves the less able to penetrate the profound abysses of his wisdom." And that this

train of thought was habitual to the philosopher
we have abundant evidence in many other parts
of his writings. He had already said in the same
dialogue, " Nature (or God, as he elsewhere
speaks) employs means in an admirable and
inconceivable manner ; admirable, that is, and
inconceivable to us, but not to her, who brings
about with consummate facility and simplicity
things which affect our intellect with infinite
astonishment. That which is to us most difficult
to understand is to her most easy to execute."

The establishment of the Copernican and
Newtonian views of the motions of the solar
system and their causes, were probably the occasions on which religious but unphilosophical men
entertained the strongest apprehensions that the
belief in the government of God may be weakened
when we thus " thrust some mechanic cause into
his place." It is therefore fortunate that we can
show, not only that this ought not to occur, from
the reason of the thing, but also that in fact the
persons who are the leading characters in the
progress of these opinions were men of clear and
fervent piety.

In the case of Copernicus himself it does not
appear that, originally, any apprehensions were
entertained of any dangerous discrepancy between
his doctrines and the truths of religion, either
natural or revealed. The work which contains
these memorable discoveries was addressed to

Pope Paul III., the head, at that time, (1543) of the religious world; and was published, as the author states in the preface, at the urgent entreaty of friends, one of whom was a cardinal, and another a bishop.* " I know," he says, " that the thoughts of a philosopher are far removed from the judgment of the vulgar; since it is his study to search out truth in all things, as far as that is permitted by God to human reason." And though the doctrines are for the most part stated as portions of a mathematical calculation, the explanation of the arrangement by which the sun is placed in the centre of the system is accompanied by a natural reflexion of a religious cast : " Who in this fair temple would place this lamp in any other or better place than there whence it may illuminate the whole? We find then under this ordination an admirable symmetry of the world, and a certain harmonious connexion of the motion and magnitude of the orbs, such as in any other way cannot be found. Thus the progressions and regressions of the planets all arise from the same cause, the motion of the earth. And that no such movements are seen in the fixed stars, argues their immense

* Amici me cunctantem atque etiam reluctantem, retraxerunt, inter quos primus fuit Nicolaus Schonbergius, Cardinalis Capuanus, in omni genere literatum celebris ; proximus ille vir mei amantissimus Tidemannus Gisius, episcopus Culmensis, sacrarum ut est et omnium bonarum literarum studiosissimus.—*De Revolutionibus. Præf. ad Paulum III.*

distance from us, which causes the apparent magnitude of the earth's annual course to become evanescent. So great, in short, is this divine fabric of the great and good God;"* "this best and most regular artificer of the universe," as he elsewhere speaks.

Kepler was the person, who by further studying "the connexion of the motions and magnitude of the orbs," to which Copernicus had thus drawn the attention of astronomers, detected the laws of this connexion, and prepared the way for the discovery, by Newton, of the mechanical laws and causes of such motions. Kepler was a man of strong and lively piety; and the exhortation which he addresses to his reader before entering on the exposition of some of his discoveries, may be quoted not only for its earnestness but its reasonableness also.—" I beseech my reader, that not unmindful of the divine goodness bestowed on man, he do with me praise and celebrate the wisdom and greatness of the Creator, which I open to him from a more inward explication of the form of the world, from a searching of causes, from a detection of the errors of vision : and that thus, not only in the firmness and stability of the earth he perceive with gratitude the preservation of all living things in nature as the gift of God, but also that in its motion, so recondite, so admirable, he acknowledge the wisdom of the

* Lib. i. cx.

Creator. But him who is too dull to receive this science, or too weak to believe the Copernican system without harm to his piety, him, I say, I advise that, leaving the school of astronomy, and condemning, if he please, any doctrines of the philosophers, he follow his own path, and desist from this wandering through the universe, and lifting up his natural eyes, with which alone he can see, pour himself out from his own heart in praise of God the Creator; being certain that he gives no less worship to God than the astronomer, to whom God has given to see more clearly with his inward eye, and who, for what he has himself discovered, both can and will glorify God."

The next great step in our knowledge of the universe, the discovery of the mechanical causes by which its motions are produced, and of their laws, has in modern times sometimes been supposed, both by the friends of religion and by others, to be unfavourable to the impression of an intelligent first cause. That such a supposition is founded in error we have offered what appear to us insurmountable reasons for believing. That in the mind of the great discoverer of this mechanical cause, Newton, the impression of a creating and presiding Deity was confirmed, not shaken, by all his discoveries, is so well known that it is almost superfluous to insist upon the fact. His views of the tendency of science invested it with no dangers of this kind. "The business of natural philosophy is," he says, (Optics, Qu. 28.) "to argue from phenomena without feigning hypotheses, and to deduce causes from effects, till we come to the very first cause, which certainly is not mechanical." "Though every true step made in this philosophy brings us not immediately to the knowledge of the first cause, yet it brings us nearer to it, and is on that account highly to be valued." The Scholium, or note, which concludes his great work, the Principia, is a well known and most striking evidence on this point, "This beautiful system of sun, planets and comets, could have its origin in no other way than by the purpose and command of an intelligent and powerful Being. He governs all things, not as the soul of the world, but as the lord of the universe. He is not only God, but Lord or Governor. We know him only by his properties and attributes, by the wise and admirable structure of things around us, and by their final causes; we admire him on account of his perfections, we venerate and worship him on account of his government."

Without making any further quotations, it must be evident to the reader that the succession of great philosophers through whom mankind have been led to the knowledge of the greatest of scientific truths, the law of universal gravitation, did, for their parts, see the truths which they disclosed to men in such a light that their religious feelings, their reference of the world to an intelligent Creator and Preserver, their admiration of his attributes, were exalted rather than impaired by the insight which they obtained into the structure of the universe.

Having shown this with regard to the most perfect portion of human knowledge, our knowledge of the motions of the solar system, we shall adduce a few other passages, illustrating the prevalence of the same fact in other departments of experimental science ; although, for reasons which have already been intimated, we conceive that sciences of experiment do not conduct so obviously as sciences of observation to the impression of a Divine Legislator of the material world.

The science of Hydrostatics was constructed in a great measure by the founders of the sister science of Mechanics. Of those who were employed in experimentally establishing the principles peculiarly belonging to the doctrine of fluids, Pascal and Boyle are two of the most eminent names. That these two great philosophers were not only religious, but both of them remarkable for their fervent and pervading devotion, is too well known to be dwelt on. With regard to Pascal, however, we ought not perhaps to pass over an opinion of his, that the existence of God cannot be proved from the external world. " I do not undertake to prove this," says he,

" not only because I do not feel myself sufficiently strong to find in nature that which shall convince obstinate atheists, but because such knowledge without Jesus Christ is useless and sterile." It is obvious that such a state of mind would prevent this writer from encouraging or dwelling upon the grounds of natural religion; while yet he himself is an example of that which we wish to illustrate, that those who have obtained the furthest insight into nature, have been in all ages firm believers in God. " Nature," he says in another place, " has perfections in order to show that she is the image of God, and defects in order to show that she is only his image."*

Boyle was not only a most pious man as well as a great philosopher, but he exerted himself very often and earnestly in his writings to show the bearing of his natural philosophy upon his views of the divine attributes, and of the government of the world. Many of these dissertations convey trains of thought and reasoning which have never been surpast for their combination of judicious sobriety in not pressing his arguments too far, with fervent devotion in his conceptions of the Divine nature. As examples of these merits, we might adduce almost any portion of his tracts on these subjects; for instance, his " Inquiry into the Final Causes of Natural

* Pensées, Art. viii. 1.

Things;" his " Free Inquiry into the Vulgar Notion of Nature ;". his " Christian Virtuoso ;" and his Essay entitled " The High Veneration Man's Intellect owes to God." It would be superflous to quote at any length from these works. We may observe, however, that he notices that general fact which we are at present employed in exemplifying, that " in almost all ages and countries the generality of philosophers and contemplative men were persuaded of the existence of a Deity from the consideration of the phenomena of the universe ; whose fabric and conduct they rationally concluded could not justly be ascribed either to chance or to any other cause than a Divine Being." And in speaking of the religious uses of science, he says : " Though I am willing to grant that some impressions of God's wisdom are so conspicuous that even a superficial philosopher may thence infer that the author of such works must be a wise agent ; yet how wise an agent he has in these works expressed himself to be, none but an experimental philosopher can well discern. And 'tis not by a slight survey, but by a diligent and skilful scrutiny, of the works of God, that a man must be, by a rational and affective conviction, engaged to acknowledge that the author of nature ' is wonderful in counsel, and excellent in working.' "

After the mechanical properties of fluids, the laws of the operation of the chemical and physical properties of the elements about us, offer themselves to our notice. The relations of heat and of moisture in particular, which play so important a part, as we have seen, in the economy of our world, have been the subject of various researches; and they have led to views of the operation of such agents, some of which we have endeavoured to present to the reader, and to point out the remarkable arrangements by which their beneficial operation is carried on. That the discoverers of the laws by which such operations are regulated, were not insensible to the persuasion of a Divine care and contrivance which those arrangements suggest, is what we should expect, in agreement with what we have already said, and it is what we find. Among the names of the philosophers to whom we owe our knowledge on these subjects, there are none greater than those of Black, the discoverer of the laws of latent heat, and Dalton, who first gave us a true view of the mode in which watery vapour exists and operates in the atmosphere. With regard to the former of these philosophers, we shall quote Dr. Thomson's account of the views which the laws of latent heat suggested to the discoverer.* " Dr. Black quickly perceived the vast importance of this discovery, and took a pleasure in laying before his students a view of

* Thomson's Hist. of Chemistry, vol. i. 321.

the beneficial effects of this habitude of heat in the economy of nature. During the summer season a vast magazine of heat is accumulated in the water, which by gradually emerging during congelation serves to temper the cold of winter. Were it not for this accumulation of heat in water and other bodies, the sun would no sooner go a few degrees to the south of the equator than we should feel all the horrors of winter."

In the same spirit are Mr. Dalton's reflexions, after pointing out the laws which regulate the balance of evaporation and rain,* which he himself first clearly explained. "It is scarcely possible," says he, "to contemplate without admiration the beautiful system of nature by which the surface of the earth is continually supplied with water, and that unceasing circulation of a fluid so essentially necessary to the very being of the animal and vegetable kingdom takes place."

Such impressions appear thus to rise irresistibly in the breasts of men, when they obtain a sight, for the first time, of the varied play and comprehensive connexions of the laws by which the business of the material world is carried on and its occurrences are brought to pass. To dwell upon or develope such reflexions is not here our business. Their general prevalence in the minds

---

* Manch. Mem. vol. v. p. 346.

W.   Y

of those to whom these first views of new truths
are granted, has been, we trust, sufficiently
illustrated. Nor are the names adduced above,
distinguished as they are, brought forwards as
authorities merely. We do not claim for the
greatest discoverers in the realms of science any
immunity from error. In their general opinions
they may, as others may, judge or reason ill.
The articles of their religious belief may be as
easily and as widely as those of other men, imperfect, perverted, unprofitable. But on this one
point, the tendency of our advances in the scientific knowledge of the universe to lead us up to a
belief in a most wise maker and master of the
universe, we conceive that they who make these
advances, and who feel, as an original impression, that which others feel only by receiving
their teaching, must be looked to with a peculiar
attention and respect. And what their impressions have commonly been, we have thus endeavoured to show.

Chapter VI.

On Deductive Habits; or, on the Impression produced on Men's Minds by tracing the consequences of ascertained Laws.

THE opinion illustrated in the last chapter, that the advances which men make in science tend to impress upon them the reality of the Divine government of the world, has often been controverted. Complaints have been made, and especially of late years, that the growth of piety has not always been commensurate with the growth of knowledge, in the minds of those who make nature their study. Views of an irreligious character have been entertained, it is sometimes said, by persons eminently well instructed in all the discoveries of modern times, no less than by the superficial and ignorant. Those who have been supposed to deny or to doubt the existence, the providence, the attributes of God, have in many cases been men of considerable eminence and celebrity for their attainments in science. The opinion that this is the case, appears to be extensively diffused, and this persuasion has probably often produced inquietude and grief in the breasts of pious and benevolent men.

This opinion, concerning the want of religious

convictions among those who have made natural philosophy their leading pursuit, has probably gone far beyond the limits of the real fact. But if we allow that there are any strong cases to countenance such an opinion, it may be worth our while to consider how far they admit of any satisfactory explanation. The fact appears at first sight to be at variance with the view we have given of the impression produced by scientific discovery; and it is moreover always a matter of uneasiness and regret, to have men of eminent talents and knowledge opposed to doctrines which we consider as important truths.

We conceive that an explanation of such cases, if they should occur, may be found in a very curious and important circumstance belonging to the process by which our physical sciences are formed. The first discovery of new general truths, and the developement of these truths when once obtained, are two operations extremely different—imply different mental habits, and may easily be associated with different views and convictions on points out of the reach of scientific demonstration. There would therefore be nothing surprising, or inconsistent with what we have maintained above, if it should appear that while original discoverers of laws of nature are peculiarly led, as we have seen, to believe the existence of a supreme intelligence and purpose; the far greater number of cultivators of science, whose employment it is to learn from others these general laws, and to trace, combine, and apply their consequences, should have no clearness of conviction or security from error on this subject, beyond what belongs to persons of any other class.

This will, perhaps, become somewhat more evident by considering a little more closely the distinction of the two operations of discovery and developement, of which we have spoken above, and the tendency which the habitual prosecution of them may be expected to produce in the thoughts and views of the student.

We have already endeavoured in some measure to describe that which takes place when a new law of nature is discovered. A number of facts in which, before, order and connexion did not appear at all, or appeared by partial and contradictory glimpses, are brought into a point of view in which order and connexion become their essential character. It is seen that each fact is but a different manifestation of the same principle; that each particular is that which it is, in virtue of the same general truth. The inscription is decyphered; the enigma is guessed; the principle is understood; the truth is enunciated.

When this step is once made, it becomes possible to deduce from the truth thus established, a train of consequences often in no small degree long and complex. The process of making these inferences may properly be described by the word Deduction. On the other hand, the very different process by which a new principle is collected from an assemblage of facts, has been termed Induction; the truths so obtained and their consequences constitute the results of the Inductive Philosophy; which is frequently and rightly described as a science which ascends from particular facts to general principles, and then descends again from these general principles to particular applications and exemplifications.

While the great and important labours by which science is really advanced consist in the successive steps of the *inductive* ascent, in the discovery of new laws perpetually more and more general; by far the greater part of our books of physical science unavoidably consists in *deductive* reasoning, exhibiting the consequences and applications of the laws which have been discovered; and the greater part of writers upon science have their minds employed in this process of deduction and application.

This is true of many of those who are considered, and justly, as distinguished and profound philosophers. In the mechanical philosophy, that science which applies the properties of matter and the laws of motion to the explanation of the phenomena of the world, this is peculiarly the case. The laws, when once discovered, occupy little room in their statement, and when no longer contested, are not felt to need a lengthened proof. But their consequences require far more room and far more intellectual labour. If we take, for example, the laws of motion and the law of universal gravitation, we can express in a few lines, that which, when developed, represents and explains an innumerable mass of natural phenomena. But here the course of developement is necessarily so long, the reasoning contains so many steps, the considerations on which it rests are so minute and refined, the complication of cases and of consequences is so vast, and even the involution arising from the properties of space and number is so serious, that the most consummate subtlety, the most active invention, the most tenacious power of inference, the widest spirit of combination, must be tasked and tasked severely, in order to solve the problems which belong to this portion of science. And the persons who have been employed on these problems, and who have brought to them the high and admirable qualities which such an office requires, have justly excited in a very eminent degree the admiration which mankind feel for great intellectual powers. Their names occupy a distinguished place in literary history ; and probably there are no scientific reputations of the last century higher, and none more merited, than those earned by the great mathematicians who have laboured with such wonderful success in unfolding the mechanism of the heavens; such for instance as D'Alembert, Clairault, Euler, Lagrange, Laplace.

But it is still important to recollect, that the mental employments of men, while they are occupied in this portion of the task of the formation of science, are altogether different from that which takes place in the mind of a discoverer, who, for the first time, seizes the principle which connects phenomena before unexplained, and thus adds another original truth to our knowledge of the universe. In explaining, as the great mathematicians just mentioned have done, the phenomena of the solar system by means of the law of universal gravitation, the conclusions at which they arrived were really included in the truth of the law, whatever skill and sagacity it might require to develope and extricate them from the general principle. But when Newton conceived and established the law itself, he added to our knowledge something which was not contained in any truth previously known, nor deducible from it by any course of mere reasoning. And the same distinction, in all other cases, obtains, between these processes which establish the principles, generally few and simple, on which our sciences rest, and those reasonings and calculations, founded on the principles thus obtained, which constitute by far the larger portion of the common treatises on the most complete of the sciences now cultivated.

Since the difference is so great between the process of inductive generalization of physical facts, and that of mathematical deduction of consequences, it is not surprising that the two processes should imply different mental powers and habits. However rare the mathematical talent, in its highest excellence, may be, it is far more common, if we are to judge from the history of science, than the genius which divines the general laws of nature. We have several good mathematicians in every age; we have few great discoverers in the whole history of our species.

The distinction being thus clearly established between original discovery and derivative speculation, between the ascent to principles and the descent from them, we have further to observe, that the habitual and exclusive prosecution of the latter process may sometimes exercise an unfavourable effect on the mind of the student, and may make him less fitted and ready to apprehend and accept truths different from those with which his reasonings are concerned. We conceive, for example, that a person labours under gross error, who believes the phenomena of the world to be altogether produced by mechanical causes, and who excludes from his view all reference to an intelligent First Cause and Governor. But we conceive that reasons may be shown which make it more probable that error of such a kind should find a place in the mind of a person of deductive, than of inductive habits;—of a mere mathematician or logician, than of one who studies the facts of the natural world and detects their laws.

The person whose mind is employed in reducing to law and order and intelligible cause the complex facts of the material world, is compelled to look beyond the present state of his knowledge, and to turn his thoughts to the existence of principles higher than those which he yet possesses. He has seen occasions when facts that at first seemed incoherent and anomalous, were reduced to rule and connexion; and when limited rules were discovered to be included in some rule of superior generality. He knows that all facts and appearances, all partial laws, however confused and casual they at present seem, must still, in reality, have this same kind of bearing and dependence;—must be bound together by some undiscovered principle of order; —must proceed from some cause working by most steady rules;—must be included in some wide and fruitful general truth. He cannot therefore consider any principles which he has already obtained, as the ultimate and sufficient reason of that which he sees. There must be some higher principle, some ulterior reason. The effort and struggle by which he endeavours to extend his view, makes him feel that there is a region of truth not included in his present physical knowledge; the very imperfection of the light in which he works his way, suggests to him that there must be a source of clearer illumination at a distance from him.

We must allow that it is scarcely possible to describe in a manner free from some vagueness and obscurity, the effect thus produced upon the mind by the efforts which it makes to reduce natural phenomena to general laws. But we trust it will still be allowed that there is no difficulty in seeing clearly that a different influence may result from this process, and from the process of deductive reasoning which forms the main employment of the mathematical cultivators and systematic expositors of physical science in modern times. Such persons are not led by their pursuits to anything beyond the general principles, which form the basis of their explanations and applications. They acquiesce in these; they make these their ultimate grounds of truth; and they are entirely employed in unfolding the particular truths which are involved in such general truths. Their thoughts dwell little upon the possibility of the laws of nature being other than we find them to be, or on the reasons why they are not so; and still less on those facts and phenomena which philosophers have not yet reduced to any rule; which are lawless to us, though we know that, in reality, they must be governed by some principle of order and harmony.

On the contrary, by assuming perpetually the existing laws as the basis of their reasoning, without question or doubt, and by employing such language that these laws can be expressed in the simplest and briefest form, they are led to think and believe as if these laws were necessarily and inevitably what they are. Some mathematicians indeed have maintained that the highest laws of nature with which we are acquainted, the laws of motion and the law of universal gravitation, are not only necessarily true, but are even self-evident and certain *a priori*, like the truths of geometry. And though the mathematical cultivator of the science of mechanics may not adopt this as his speculative opinion, he may still be so far influenced by the tendency from which it springs, that he may rest in the mechanical laws of the universe as ultimate and all-sufficient principles, without seeing in them any evidence of their having been selected and ordained, and thus without ascending from the contemplation of the world to the thought of an Intelligent Ruler. He may thus substitute for the Deity certain axioms and first principles, as the cause of all. And the follower of Newton may run into the error with which he is sometimes charged, of thrusting some mechanic cause into the place of God, if he do not raise his views, as his master did, to some higher cause, to some source of all forces, laws, and principles.

When, therefore, we consider the mathematicians who are employed in successfully applying the mechanical philosophy, as men well deserving of honour from those who take an interest in the progress of science, we do rightly; but it is still to be recollected, that in doing this they are not carrying us to any higher point of view in the knowledge of nature than we had attained before : they are only unfolding the consequences, which were already virtually in our possession, because they were implied in principles already discovered :—they are adding to our knowledge of effects, but not to our knowledge of causes :—they are not making any advance in that progress of which Newton spoke, and in which he made so vast a stride, in which " every step made brings us nearer to the knowledge of the first cause, and is on that account highly to be valued." And as in this advance they have no peculiar privileges or advantages, their errors of opinion concerning it, if they err, are no more to be wondered at, than those of common men ; and need as little disturb or distress us, as if those who committed them had confined themselves to the study of arithmetic or of geometry. If we can console and tranquillize ourselves concerning the defective or perverted views of religious truth entertained by any of our fellow men, we need find no additional difficulty in doing so when those who are mistaken are great mathematicians, who have added to the riches and elegance of the mechanical philosophy. And if we are seeking for extraneous grounds of trust and comfort on this subject, we may find them in the reflexion; —that, whatever may be the opinions of those who assume the causes and laws of that philosophy and reason from them, the views of those admirable and ever-honoured men who first caught sight of these laws and causes, impressed *them* with the belief that this is " the fabric of a great and good God;" that " it is man's duty to pour out his soul in praise of the Creator;" and that all this beautiful system must be referred to " a first cause, which is certainly not mechanical."

2. We may thus, with the greatest propriety, deny to the mechanical philosophers and mathematicians of recent times any authority with regard to their views of the administration of the universe; we have no reason whatever to expect from their speculations any help, when we attempt to ascend to the first cause and supreme ruler of the universe. But we might perhaps go further, and assert that they are in some respects less likely than men employed in other pursuits, to make any clear advance towards such a subject of speculation. Persons whose thoughts are thus entirely occupied in deduction are apt to forget that this is, after all, only one employment of the reason among more; only one mode of arriving at truth, needing to have its deficiencies com-

pleted by another. Deductive reasoners, those who cultivate science of whatever kind, by means of mathematical and logical processes alone, may acquire an exaggerated feeling of the amount and value of their labours. Such employments, from the clearness of the notions involved in them, the irresistible concatenation of truths which they unfold, the subtlety which they require, and their entire success in that which they attempt, possess a peculiar fascination for the intellect. Those who pursue such studies have generally a contempt and impatience of the pretensions of all those other portions of our knowledge, where from the nature of the case, or the small progress hitherto made in their cultivation, a more vague and loose kind of reasoning seems to be adopted. Now if this feeling be carried so far as to make the reasoner suppose that these mathematical and logical processes can lead him to all the knowledge and all the certainty which we need, it is clearly a delusive feeling. For it is confessed on all hands, that all which mathematics or which logic can do, is to develope and extract those truths, as conclusions, which were in reality involved in the principles on which our reasonings proceeded.* And this being allowed, we cannot but ask how we obtain

* " Since all reasoning may be resolved into syllogisms, and since in a syllogism the premises do virtually assert the conclu-

these principles? from what other source of
knowledge we derive the original truths which
we thus pursue into detail? since it is manifest
that such principles cannot be derived from the
proper stores of mathematics or logic. These
methods can generate no new truth; and all the
grounds and elements of the knowledge which,
through them, we can acquire, must necessarily
come from some extraneous source. It is certain, therefore, that the mathematician and the
logician must derive from some process different
from their own, the substance and material of all
our knowledge, whether physical or metaphysical, physiological or moral. This process, by
which we acquire our first principles, (without
pretending here to analyse it,) is obviously the
general course of human experience, and the
natural exercise of the understanding; our intercourse with matter and with men, and the consequent growth in our minds of convictions and
conceptions such as our reason can deal with,
either by her systematic or unsystematic methods
of procedure. Supplies from this vast and inexhaustible source of original truths are requisite,
to give any value whatever to the results of our
deductive processes, whether mathematical or

sion, it follows at once, that no new truth can be elicited by any
process of reasoning.— *Whately's Logic*, p. 223.

Mathematics is the *logic of quantity*, and to this science the
observation here quoted is strictly applicable.

logical; while, on the other hand, there are many branches of our knowledge, in which we possess a large share of original and derivative convictions and truths, but where it is nevertheless at present quite impossible to erect our knowledge into a complete system ;—to state our primary and independent truths, and to show how on these all the rest depend by the rules of art. If the mathematician is repelled from speculations on morals or politics, on the beautiful or the right, because the reasonings which they involve have not mathematical precision and conclusiveness, he will remain destitute of much of the most valuable knowledge which man can acquire. And if he attempts to mend the matter by giving to treatises on morals, or politics, or criticism, a form and a phraseology borrowed from the very few tolerably complete physical sciences which exist, it will be found that he is compelled to distort and damage the most important truths, so as to deprive them of their true shape and import, in order to force them into their places in his artificial system.

If therefore, as we have said, the mathematical philosopher dwells in his own bright and pleasant land of deductive reasoning, till he turns with disgust from all the speculations, necessarily less clear and conclusive, in which his imagination, his practical faculties, his moral sense, his capacity of religious hope and belief, are to be

W.   Z

called into action, he becomes, more than common men, liable to miss the roads to truths of extreme consequence.

This is so obvious, that charges are frequently brought against the study of mathematics, as unfitting men for those occupations which depend upon our common instinctive convictions and feelings, upon the unsystematic exercise of the understanding with regard to common relations and common occurrences. Bonaparte observed of Laplace, when he was placed in a public office of considerable importance, that he did not discharge it in so judicious and clear-sighted a manner as his high intellectual fame might lead most persons to expect.* " He sought," that great judge of character said, " subtleties in every subject, and carried into his official employments the spirit of the method of infinitely small quantities," by which the mathematician solves his most abstruse problems. And the complaint that mathematical studies make men insensible to moral evidence and to poetical beauties, is so often repeated as to show that

* A l'intérieur le ministre Quinette fut remplacé par Laplace, géomêtre du premier rang, mais qui ne tarda pas à se montrer administrateur plus que médiocre: des son premier travail les consuls s'aperçurent qu'ils s'étaient trompés : Laplace ne saisissait aucune question sous son vrai point de vue : il cherchait des subtilités partout, n'avait que des idées problématiques, et portait enfin l'esprit des infiniment petits dans l'administration.—*Mémoires écrits à Ste Hélène,* i. 3.

some opposition of tendency is commonly perceived between that exercise of the intellect which mathematics requires and those processes which go on in our minds when moral character or imaginative beauty is the subject of our contemplation.

Thus, while we acknowledge all the beauty and all the value of the mathematical reasonings by which the consequences of our general laws are deduced, we may yet consider it possible that a philosopher, whose mind has been mainly employed, and his intellectual habits determined, by this process of deduction, may possess, in a feeble and imperfect degree only, some of those faculties by which truth is attained, and especially truths such as regard our relation to that mind, which is the origin of all law, the source of first principles, and which must be immeasurably elevated above all derivative truths. It would, therefore, be far from surprising, if there should be found, among the great authors of the developements of the mechanical philosophy, some who had refused to refer the phenomena of the universe to a supreme mind, purpose, and will. And though this would be, to a believer in the Being and government of God, a matter of sorrow and pain, it need not excite more surprise than if the same were true of a person of the most ordinary endowments, when it is recollected in what a disproportionate manner the various faculties of such a philosopher may have been cultivated. And our apprehensions of injury to mankind from the influence of such examples will diminish, when we consider, that those mathematicians whose minds have been less partially exercised, the great discoverers of the truths which others apply, the philosophers who have looked upwards as well as downwards, to the unknown as well as to the known, to ulterior as well as proximate principles, have never rested in this narrow and barren doctrine; but have perpetually extended their view forwards, beyond mere material laws and causes, to a First Cause of the moral and material world, to which each advance in philosophy might bring them nearer, though its highest attributes must probably ever remain indefinitely beyond their reach.

It scarcely needs, perhaps, to be noticed, that what we here represent as the possible source of error is, not the perfection of the mathematical habits of the mind, but the deficiency of the habit of apprehending truth of other kinds;—not a clear insight into the mathematical consequences of principles, but a want of a clear view of the nature and foundation of principles;—not the talent for generalizing geometrical or mechanical relations, but the tendency to erect such relations into ultimate truths and efficient causes. The most consummate mathematical skill may accompany and be auxiliary to the most earnest piety, as it often has been. And an entire command of the conceptions and processes of mathematics is not only consistent with, but is the necessary condition and principal instrument of every important step in the discovery of physical principles. Newton was eminent above the philosophers of his time, in no one talent so much as in the power of mathematical deduction. When he had caught sight of the law of universal gravitation, he traced it to its consequences with a rapidity, a dexterity, a beauty of mathematical reasoning which no other person could approach ; so that on this account, if there had been no other, the establishment of the general law was possible to him alone. He still stands at the head of mathematicians as well as of philosophical discoverers. But it never appeared to him, as it may have appeared to some mathematicians who have employed themselves on his discoveries, that the general law was an ultimate and sufficient principle ; that the point to which he had hung his chain of deduction was the highest point in the universe. Lagrange, a modern mathematician of transcendent genius, was in the habit of saying, in his aspirations after future fame, that Newton was fortunate in having had the system of the world for his problem, since its theory could be discovered once only. But Newton himself appears to have had no such persuasion that the problem he had solved was unique and final ; he laboured to reduce gravity to some higher law, and the forces of other physical operations to an analogy with those of gravity, and declared that all these were but steps in our advance towards a first cause. Between us and this first cause, the source of the universe and of its laws, we cannot doubt that there intervene many successive steps of possible discovery and generalization, not less wide and striking than the discovery of universal gravitation : but it is still more certain that no extent or success of physical investigation can carry us to any point which is not at an immeasurable distance from an adequate knowledge of Him.

ON THE

PHILOSOPHY

OF

DISCOVERY

CHAPTER XXII.

Mr. Mill's Logic[1].

THE *History of the Inductive Sciences* was published in 1837, and the *Philosophy of the Inductive Sciences* in 1840. In 1843 Mr. Mill published his
System of Logic, in which he states that without the
aid derived from the facts and ideas in my volumes,
the corresponding portion of his own would most probably not have been written, and quotes parts of what
I have said with commendation. He also, however,
dissents from me on several important and fundamental points, and argues against what I have said
thereon. I conceive that it may tend to bring into a
clearer light the doctrines which I have tried to establish, and the truth of them, if I discuss some of the
differences between us, which I shall proceed to do[2].

Mr. Mill's work has had, for a work of its abstruse
character, a circulation so extensive, and admirers so
numerous and so fervent, that it needs no commendation of mine. But if my main concern at present had
not been with the points in which Mr. Mill *differs*
from me, I should have had great pleasure in pointing
out passages, of which there are many, in which Mr.
Mill appears to me to have been very happy in pro-
moting or in expressing philosophical truth.

There is one portion of his work indeed which
tends to give it an interest of a wider kind than be-

---

[1] *A System of Logic, Ratiocinative
and Inductive, being a connected view
of the Principles of Evidence, and of
the Methods of Scientific Investigation.* By John Stuart Mill.]

[2] These Remarks were published
in 1849, under the title *Of Induction,
with especial reference to Mr. J.
Mill's System of Logic.*

longs to that merely scientific truth to which I purposely and resolutely confined my speculations in the
works to which I have referred. Mr. Mill has introduced into his work a direct and extensive consideration of the modes of dealing with moral and political
as well as physical questions; and I have no doubt
that this part of his book has, for many of his readers,
a more lively interest than any other. Such a comprehensive scheme seems to give to doctrines respecting science a value and a purpose which they cannot
have, so long as they are restricted to mere material
sciences. I still retain the opinion, however, upon
which I formerly acted, that the philosophy of science
is to be extracted from the portions of science which
are universally allowed to be most certainly established, and that those are the physical sciences. I am
very far from saying, or thinking, that there is no
such thing as Moral and Political Science, or that no
method can be suggested for its promotion; but I
think that by attempting at present to include the
Moral Sciences in the same formulæ with the Physical, we open far more controversies than we close;
and that in the moral as in the physical sciences, the
first step towards showing how truth is to be discovered, is to study some portion of it which is assented
to so as to be beyond controversy.

I. *What is Induction?*—1. Confining myself, then,
to the material sciences, I shall proceed to offer my
remarks on Induction with especial reference to Mr.
Mill's work. And in order that we may, as I have
said, proceed as intelligibly as possible, let us begin
by considering what we mean by *Induction*, as a mode
of obtaining truth; and let us note whether there is
any difference between Mr. Mill and me on this subject.

" For the purposes of the present inquiry," Mr. Mill
says (i. 347[3]), "Induction may be defined the opera-

---

[3] My references are throughout the volume and the page of Mr. Mill's
(except when otherwise expressed) to first edition of his *Logic.*

tion of discovering and forming general propositions:" meaning, as appears by the context, the discovery of them from particular facts. He elsewhere (i. 370) terms it "generalization from experience:" and again he speaks of it with greater precision as the inference of a more general proposition from less general ones.

2. Now to these definitions and descriptions I assent as far as they go; though, as I shall have to remark, they appear to me to leave unnoticed a feature which is very important, and which occurs in all cases of Induction, so far as we are concerned with it. Science, then, consists of general propositions, inferred from particular facts, or from less general propositions, by Induction; and it is our object to discern the nature and laws of *Induction* in this sense. That the propositions are general, or are more general than the facts from which they are inferred, is an indispensable part of the notion of Induction, and is essential to any discussion of the process, as the mode of arriving at Science, that is, at a body of general truths.

3. I am obliged therefore to dissent from Mr. Mill when he includes, in his notion of Induction, the process by which we arrive *at individual facts* from other facts *of the same order of particularity.*

Such inference is, at any rate, not Induction *alone;* if it be Induction at all, it is Induction applied to an example.

For instance, it is a general law, obtained by Induction from particular facts, that a body falling vertically downwards from rest, describes spaces proportional to the squares of the times. But that a particular body will fall through 16 feet in one second and 64 feet in two seconds, is not an induction simply, it is a result obtained by applying the inductive law to a particular case.

But further, such a process is often not induction *at all.* That a ball striking another ball directly will communicate to it as much momentum as the striking ball itself loses, is a law established by induction: but if, from habit or practical skill, I make one billiard-ball strike another, so as to produce the velocity which

I wish, without knowing or thinking of the general law, the term *Induction* cannot then be rightly applied. If I *know the law* and act upon it, I have in my mind both the general induction and its particular application. But if I act by the ordinary billiard-player's skill, without thinking of momentum or law, there is no Induction in the case.

II. *Induction or Description?*—15. In the cases hitherto noticed, Mr. Mill extends the term *Induction*, as I think, too widely, and applies it to cases to which it is not rightly applicable. I have now to notice a case of an opposite kind, in which he does not apply it where I do, and condemns me for using it in such a case. I had spoken of Kepler's discovery of the Law, that the planets move round the sun in ellipses, as an example of Induction. The separate facts of any planet (Mars, for instance,) being in certain places at certain times, are all included in the general proposition which Kepler discovered, that Mars describes an ellipse of a certain form and position. This appears to me a very simple but a very distinct example of the operation of discovering general propositions; general, that is, with reference to particular facts; which operation Mr. Mill, as well as myself, says is Induction. But Mr. Mill denies this operation in this case to be Induction at all (i. 357). I should not have been prepared for this denial by the previous parts of Mr. Mill's book, for he had said just before (i. 350), "such facts as the magnitudes of the bodies of the solar system, their distances from each other, the figure of the earth and its rotation...are proved indirectly, by the aid of inductions founded on other facts which we can more easily reach." If the figure of the earth and its rotation are proved by Induction, it seems very strange, and is to me quite incomprehensible, how the figure of the earth's orbit and its revolution (and of course, of the figure of Mars's orbit and his revolution in like manner,) are not also proved by Induction. No, says Mr. Mill, Kepler, in putting together a number of places of the planet into one figure, only performed an act of *description*. "This descriptive operation," he adds (i. 359), "Mr. Whewell, by an aptly chosen expression, has termed Colligation of Facts." He goes on to commend my observations concerning this process, but says that, according to the old and received meaning of the term, it is not Induction at all.

16. Now I have already shown that Mr. Mill himself, a few pages earlier, had applied the term *Induction* to cases undistinguishable from this in any essential circumstance. And even in this case, he allows that Kepler did really perform an act of Induction (i. 358), "namely, in concluding that, because the observed places of Mars were correctly represented by points in an imaginary ellipse, therefore Mars would continue to revolve in that same ellipse; and even in concluding that the position of the planet during the time which had intervened between the two observations must have coincided with the intermediate points of the curve." Of course, in Kepler's Induction, of which I speak, I include all this; all this is included in speaking of the *orbit* of Mars : a continuous line, a periodical motion, are implied in the term *orbit.* I am unable to see what would remain of Kepler's discovery, if we take from it these conditions. It would not only not be an induction, but it would not be a description, for it would not recognize that Mars moved in an orbit. Are particular positions to be conceived as points in a curve, without thinking of the intermediate positions as belonging to the same curve? If so, there is no law at all, and the facts are not bound together by any intelligible tie.

In another place (ii. 209) Mr. Mill returns to his distinction of Description and Induction; but without throwing any additional light upon it, so far as I can see.

17. The only meaning which I can discover in this attempted distinction of Description and Induction is, that when particular facts are bound together by their relation in *space*, Mr. Mill calls the discovery of the connexion *Description*, but when they are connected by other general relations, as time, cause and the like, Mr. Mill terms the discovery of the connexion *Induc-*

tion. And this way of making a distinction, would
fall in with the doctrine of other parts of Mr. Mill's
book, in which he ascribes very peculiar attributes to
space and its relations, in comparison with other Ideas,
(as I should call them). But I cannot see any ground
for this distinction, of connexion according to space
and other connexions of facts.

To stand upon such a distinction, appears to me to
be the way to miss the general laws of the formation
of science. For example: The ancients discovered
that the planets revolved in recurring periods, and
thus connected the observations of their motions according to the Idea of *Time.* Kepler discovered that
they revolved in ellipses, and thus connected the observations according to the Idea of *Space.* Newton
discovered that they revolved in virtue of the Sun's
attraction, and thus connected the motions according
to the Idea of *Force.* The first and third of these discoveries are recognized on all hands as processes of
Induction. Why is the second to be called by a different name? or what but confusion and perplexity
can arise from refusing to class it with the other two?
It is, you say, Description. But such Description is a
kind of Induction, and must be spoken of as Induction,
if we are to speak of Induction as the process by which
Science is formed: for the three steps are all, the
second in the same sense as the first and third, in
co-ordination with them, steps in the formation of
astronomical science.

18. But, says Mr. Mill (i. 363), "it is a fact surely
that the planet does describe an ellipse, and a fact
which we could see if we had adequate visual organs
and a suitable position." To this I should reply: "Let
it be so; and it is a fact, surely, that the planet does
move periodically: it is a fact, surely, that the planet
is attracted by the sun. Still, therefore, the asserted
distinction fails to find a ground." Perhaps Mr. Mill
would remind us that the elliptical form of the orbit is
a fact which we could see if we had adequate visual
organs and a suitable position: but that force is a
thing which we cannot see. But this distinction also

will not bear handling. Can we not see a tree blown down by a storm, or a rock blown up by gunpowder? Do we not here see force :—see it, that is, by its effects, the only way in which we need to see it in the case of a planet, for the purposes of our argument? Are not such operations of force, Facts which may be the objects of sense? and is not the operation of the sun's Force a Fact of the same kind, just as much as the elliptical form of orbit which results from the action? If the latter be "surely a Fact," the former is a Fact no less surely.

19. In truth, as I have repeatedly had occasion to remark, all attempts to frame an argument by the exclusive or emphatic appropriation of the term *Fact* to particular cases, are necessarily illusory and inconclusive. There is no definite and stable distinction between Facts and Theories; Facts and Laws; Facts and Inductions. Inductions, Laws, Theories, which are true, *are* Facts. Facts involve Inductions. It is a fact that the moon is attracted by the earth, just as much as it is a Fact that an apple falls from a tree. That the former fact is collected by a more distinct and conscious Induction, does not make it the less a Fact. That the orbit of Mars is a Fact—a true Description of the path—does not make it the less a case of Induction.

20. There is another argument which Mr. Mill employs in order to show that there is a difference between mere colligation which is description, and induction in the more proper sense of the term. He notices with commendation a remark which I had made (i. 364), that at different stages of the progress of science the facts had been successfully connected by means of very different conceptions, while yet the later conceptions have not contradicted, but included, so far as they were true, the earlier: thus the ancient Greek representation of the motions of the planets by means of epicycles and eccentrics, was to a certain degree of accuracy true, and is not negatived, though superseded, by the modern representation of the planets as describing ellipses round the sun. And he then reasons that

this, which is thus true of Descriptions, cannot be true of Inductions. He says (i. 367), "Different descriptions therefore may be all true: but surely not different explanations." He then notices the various explanations of the motions of the planets—the ancient doctrine that they are moved by an inherent virtue; the Cartesian doctrine that they are moved by impulse and by vortices; the Newtonian doctrine that they are governed by a central force; and he adds, "Can it be said of these, as was said of the different descriptions, that they are all true as far as they go? Is it not true that one only can be true in any degree, and that the other two must be altogether false?"

21. And to this questioning, the history of science compels me to reply very distinctly and positively, in the way which Mr. Mill appears to think extravagant and absurd. I am obliged to say, Undoubtedly, all these explanations *may* be true and consistent with each other, and would be so if each had been followed out so as to show in what manner it could be made consistent with the facts. And this was, in reality, in a great measure done[4]. The doctrine that the heavenly bodies were moved by vortices was successively modified, so that it came to coincide in its results with the doctrine of an inverse-quadratic centripetal force, as I have remarked in the *History*[5]. When this point was reached, the vortex was merely a machinery, well or ill devised, for producing such a centripetal force, and therefore did not contradict the doctrine of a centripetal force. Newton himself does not appear to have been averse to explaining gravity by impulse. So little is it true that if the one theory be true the other must be false. The attempt to explain gravity by the impulse of streams of particles flowing through the universe in all directions, which I have mentioned in the *Philosophy*[6], is so far from being incon-

---

[4] On this subject see an Essay *On the Transformation of Hypotheses*, given in the Appendix.

[5] B. vii. c. iii. sect. 3.   [6] B. iii. c. ix. art. 7.

sistent with the Newtonian theory, that it is founded entirely upon it. And even with regard to the doctrine, that the heavenly bodies move by an inherent virtue; if this doctrine had been maintained in any such way that it was brought to agree with the facts, the inherent virtue must have had its laws determined; and then, it would have been found that the virtue had a reference to the central body; and so, the " inherent virtue" must have coincided in its effect with the Newtonian force; and then, the two explanations would agree, except so far as the word "inherent" was concerned. And if such a part of an earlier theory as this word *inherent* indicates, is found to be untenable, it is of course rejected in the transition to later and more exact theories, in Inductions of this kind, as well as in what Mr. Mill calls Descriptions. There is therefore still no validity discoverable in the distinction which Mr. Mill attempts to draw between "descriptions" like Kepler's law of elliptical orbits, and other examples of induction.

22. When Mr. Mill goes on to compare what he calls different predictions—the first, the true explanation of eclipses by the shadows which the planets and satellites cast upon one another, and the other, the belief that they will occur whenever some great calamity is impending over mankind, I must reply, as I have stated already, (Art. 17), that to class such superstitions as the last with cases of Induction, appears to me to confound all use of words, and to prevent, as far as it goes, all profitable exercise of thought. What possible advantage can result from comparing (as if they were alike) the relation of two descriptions of a phenomenon, each to a certain extent true, and therefore both consistent, with the relation of a scientific truth to a false and baseless superstition ?

23. But I may make another remark on this example, so strangely introduced. If, under the influence of fear and superstition, men may make such mistakes with regard to laws of nature, as to imagine that eclipses portend calamities, are they quite secure from mistakes in *description?* Do not the very per-

sons who tell us how eclipses predict disasters, also describe to us fiery swords seen in the air, and armies fighting in the sky ? So that even in this extreme case, at the very limit of the rational exercise of human powers, there is nothing to distinguish Description from Induction.

I shall now leave the reader to judge whether this feature in the history of science,—that several views which appear at first quite different are yet all true,— which Mr. Mill calls a curious and interesting remark of mine, and which he allows to be "strikingly true" of the Inductions which he calls *Descriptions,* (i. 364) is, as he says, "unequivocally false" of other Inductions. And I shall confide in having general assent with me, when I continue to speak of Kepler's *Induction* of the elliptical orbits.

I now proceed to another remark.

37. There is no doubt that when we discover a law of nature by induction, we find some point in which all the particular facts agree. All the orbits of the planets agree in being ellipses, as Kepler discovered ; all falling bodies agree in being acted on by a uniform force, as Galileo discovered ; all refracted rays agree in having the sines of incidence and refraction in a constant ratio, as Snell discovered ; all the bodies in the universe agree in attracting each other, as Newton discovered ; all chemical compounds agree in being constituted of elements in definite proportions, as Dalton discovered. But it appears to me a most scanty, vague, and incomplete account of these steps in science, to say that the authors of them discovered something in which the facts in each case agreed. The point in which the cases agree, is of the most diverse kind in the different cases—in some, a relation of space, in others, the action of a force, in others, the mode of composition of a substance;—and the point of agreement, visible to the discoverer alone, does not come even into his sight, till after the facts have been connected by thoughts of his own, and regarded in

points of view in which he, by his mental acts, places them. It would seem to me not much more inappropriate to say, that an officer, who disciplines his men till they move together at the word of command, does so by finding something in which they agree. If the power of consentaneous motion did not exist in the individuals, he could not create it : but that power being there, he finds it and uses it. Of course I am aware that the parallel of the two cases is not exact; but in the one case, as in the other, that in which the particular things are found to agree, is something formed in the mind of him who brings the agreement into view.

IV. *Mr. Mill's Four Methods of Inquiry.*—38. Mr. Mill has not only thus described the business of scientific discovery; he has also given rules for it, founded on this description. It may be expected that we should bestow some attention upon the methods of inquiry which he thus proposes. I presume that they are regarded by his admirers as among the most valuable parts of his book; as certainly they cannot fail to be, if they describe methods of scientific inquiry in such a manner as to be of use to the inquirer.

Mr. Mill enjoins four methods of experimental inquiry, which he calls *the Method of Agreement, the Method of Difference, the Method of Residues,* and *the Method of Concomitant Variations*[8]. They are all described by formulæ of this kind:—Let there be, in the observed facts, combinations of antecedents, *ABC, BC, ADE,* &c. and combinations of corresponding consequents, *abc, bc, ade,* &c.; and let the object of inquiry be, the consequence of some cause *A,* or the cause of some consequence *a.* The Method of Agreement teaches us, that when we find by experiment such facts as *abc* the consequent of *ABC,* and *ade* the consequent of *ADE,* then *a* is the consequent of *A.* The Method of Difference teaches us that

---

[8] B. iii. c. viii.

when we find such facts as *abc* the consequent of *ABC*, and *bc* the consequent of *BC*, then *a* is the consèquent of *A*. The Method of Residues teaches us, that if *abc* be the consequent of *ABC*, and if we have already ascertained that the effect of *A* is *a*, and the effect of *B* is *b*, then we may infer that the effect of *C* is *c*. The Method of Concomitant Variations teaches us, that if a phenomenon *a* varies according as another phenomenon *A* varies, there is some connexion of causation direct or indirect, between *A* and *a*.

39. Upon these methods, the obvious thing to remark is, that they take for granted the very thing which is most difficult to discover, the reduction of the phenomena to formulæ such as are here presented to us. When we have any set of complex facts offered to us; for instance, those which were offered in the cases of discovery which I have mentioned,—the facts of the planetary paths, of falling bodies, of refracted rays, of cosmical motions, of chemical analysis; and when, in any of these cases, we would discover the law of nature which governs them, or, if any one chooses so to term it, the feature in which all the cases agree, where are we to look for our *A*, *B*, *C* and *a*, *b*, *c*? Nature does not present to us the cases in this form; and how are we to reduce them to this form? You say, *when* we find the combination of *ABC* with *abc* and *ABD* with *abd*, then we may draw our inference. Granted: but when and where are we to find such combinations? Even now that the discoveries are made, who will point out to us what are the *A*, *B*, *C* and *a*, *b*, *c* elements of the cases which have just been enumerated? Who will tell us which of the methods of inquiry those historically real and successful inquiries exemplify? Who will carry these formulæ through the history of the sciences, as they have really grown up; and show us that these four methods have been operative in their formation; or that any light is thrown upon the steps of their progress by reference to these formulæ?

40. Mr. Mill's four methods have a great resemblance to Bacon's "Prerogatives of Instances;" for

example, the Method of Agreement to the *Instantiæ Ostensivæ;* the Method of Differences to the *Instantiæ Absentiæ in Proximo,* and the *Instantiæ Crucis;* the Method of Concomitant Variations to the *Instantiæ Migrantes.* And with regard to the value of such methods, I believe all study of science will convince us more and more of the wisdom of the remarks which Sir John Herschel has made upon them[9].

"It has always appeared to us, we must confess, that the help which the classification of instances under their different titles of prerogative, affords to inductions, however just such classification may be in itself, is yet more apparent than real. The force of the instance must be felt in the mind before it can be referred to its place in the system; and before it can be either referred or appreciated it must be known; and when it *is* appreciated, we are ready enough to weave our web of induction, without greatly troubling ourselves whence it derives the weight we acknowledge it to have in our decisions....No doubt such instances as these are highly instructive; but the difficulty in physics is to find such, not to perceive their force when found."

V. *His Examples.*—41. If Mr. Mill's four methods had been applied by him in his book to a large body of conspicuous and undoubted examples of discovery, well selected and well analysed, extending along the whole history of science, we should have been better able to estimate the value of these methods. Mr. Mill has certainly offered a number of examples of his methods; but I hope I may say, without offence, that they appear to me to be wanting in the conditions which I have mentioned. As I have to justify myself for rejecting Mr. Mill's criticism of doctrines which I have put forward, and examples which I have adduced, I may, I trust, be allowed to offer some critical remarks in return, bearing upon the examples which he

---

[9] *Discourse,* Art. 192.

has given, in order to illustrate his doctrines and precepts.

42. The first remark which I have to make is, that a large proportion of his examples (i. 480, &c.) is taken from one favourite author; who, however great his merit may be, is too recent a writer to have had his discoveries confirmed by the corresponding investigations and searching criticisms of other labourers in the same field, and placed in their proper and permanent relation to established truths; these alleged discoveries being, at the same time, principally such as deal with the most complex and slippery portions of science, the laws of vital action. Thus Mr. Mill has adduced, as examples of discoveries, Prof. Liebig's doctrine—that death is produced by certain metallic poisons through their forming indecomposable compounds; that the effect of respiration upon the blood consists in the conversion of peroxide of iron into protoxide—that the antiseptic power of salt arises from its attraction for moisture—that chemical action is contagious; and others. Now supposing that we have no doubt of the truth of these discoveries, we must still observe that they cannot wisely be cited, in order to exemplify the nature of the progress of knowledge, till they have been verified by other chemists, and worked into their places in the general scheme of chemistry; especially, since it is tolerably certain that in the process of verification, they will be modified and more precisely defined. Nor can I think it judicious to take so large a proportion of our examples from a region of science in which, of all parts of our material knowledge, the conceptions both of ordinary persons, and even of men of science themselves, are most loose and obscure, and the genuine principles most contested; which is the case in physiology. It would be easy, I think, to point out the vague and indeterminate character of many of the expressions in which the above examples are propounded, as well as their doubtful position in the scale of chemical gene-

ralization; but I have said enough to show why I
cannot give much weight to these, as cardinal examples of the method of discovery; and therefore I shall
not examine in detail how far they support Mr. Mill's
methods of inquiry.

52. There is another kind of evidence of theories,
very closely approaching to the verification of untried
predictions, and to which, apparently, Mr. Mill does
not attach much importance, since he has borrowed
the term by which I have described it, *Consilience*,
but has applied it in a different manner (ii. 530,
563, 590). I have spoken, in the *Philosophy*[18], of
the *Consilience of Inductions*, as one of the *Tests of
Hypotheses*, and have exemplified it by many instances;
for example, the theory of universal gravitation, obtained by induction from the motions of the planets,
was found to explain also that peculiar motion of
the spheroidal earth which produces the Precession
of the Equinoxes. This, I have said, was a striking
and surprising coincidence which gave the theory a
stamp of truth beyond the power of ingenuity to
counterfeit. I may compare such occurrences to a
case of interpreting an unknown character, in which
two different inscriptions, deciphered by different
persons, had given the same alphabet. We should,
in such a case, believe with great confidence that the
alphabet was the true one; and I will add, that I
believe the history of science offers no example in
which a theory supported by such consiliences, had
been afterwards proved to be false.

53. Mr. Mill accepts (ii. 21) a rule of M. Comte's,
that we may apply hypotheses, provided they are capable of being afterwards verified as facts. I have a
much higher respect for Mr. Mill's opinion than for

[18] B. xi. c. v. art. 11.

M. Comte's[19]; but I do not think that this rule will be
found of any value. It appears to me to be tainted
with the vice which I have already noted, of throwing
the whole burthen of explanation upon the unexplained word *fact* — unexplained in any permanent
and definite opposition to theory. As I have said,
the Newtonian theory *is* a fact. Every true theory
is a fact. Nor does the distinction become more clear
by Mr. Mill's examples. "The vortices of Descartes
would have been," he says, "a perfectly legitimate
hypothesis, if it had been possible by any mode of
explanation which we could entertain the hope of
possessing, to bring the question whether such vortices
exist or not, within the reach of our observing faculties." But this was possible, and was done. The free
passage of comets through the spaces in which these
vortices should have been, convinced men that these
vortices did not exist. In like manner Mr. Mill rejects the hypothesis of a luminiferous ether, "because
it can neither be seen, heard, smelt, tasted, or touched."
It is a strange complaint to make of the vehicle of
light, that it cannot be heard, smelt, or tasted. Its
vibrations *can* be seen. The fringes of shadows for
instance, show its vibrations, just as the visible lines

---

[19] I have given elsewhere (see last
chapter) reasons why I cannot assign
to M. Comte's *Philosophie Positive*
any great value as a contribution to
the philosophy of science. In this
judgment I conceive that I am supported by the best philosophers of
our time. M. Comte owes, I think,
much of the notice which has been
given to him to his including, as Mr.
Mill does, the science of society and
of human nature in his scheme,
and to his boldness in dealing with
these. He appears to have been received with deference as a mathematician: but Sir John Herschel has
shown that a supposed astronomical
discovery of his is a mere assumption. I conceive that I have shown
that his representation of the history
of science is erroneous, both in its
details and in its generalities. His
distinction of the three stages of sciences, the theological, metaphysical,
and positive, is not at all supported
by the facts of scientific history.
Real discoveries always involve what
he calls *metaphysics;* and the doctrine of final causes in physiology,
the main element of science which
can properly be called *theological,*
is retained at the end, as well as the
beginning of the science, by all except a peculiar school.

of waves near the shore show the undulations of the
sea. Whether this can be touched, that is, whether it resists motion, is hardly yet clear. I am far
from saying there are not difficulties on this point,
with regard to *all* theories which suppose a *medium*.
But there are no more difficulties of this kind in the
undulatory theory of light, than there are in Fourier's
theory of heat, which M. Comte adopts as a model of
scientific investigation; or in the theory of voltaic
currents, about which Mr. Mill appears to have no
doubt; or of electric *atmospheres*, which, though generally obsolete, Mr. Mill appears to favour; for though
it had been said that we *feel* such atmospheres, no one
had said that they have the other attributes of matter.

VIII. *Newton's Vera Causa.*—54. Mr. Mill conceives (ii. 17) that his own rule concerning hypotheses
coincides with Newton's Rule, that the cause assumed
must be a *vera causa*. But he allows that "Mr.
Whewell...has had little difficulty in showing that his
(Newton's) conception was neither precise nor consistent with itself." He also allows that "Mr. Whewell
is clearly right in denying it to be necessary that
the cause assigned should be a cause already known;
else how could we ever become acquainted with new
causes?" These points being agreed upon, I think that
a little further consideration will lead to the conviction
that Newton's Rule of philosophizing will best become
a valuable guide, if we understand it as asserting that
when the explanation of two or more different kinds
of phenomena (as the revolutions of the planets, the
fall of a stone, and the precession of the equinoxes,)
lead us to *the same* cause, such a coincidence gives a
reality to the cause. We have, in fact, in such a case,
a Consilience of Inductions.

74. I have now made such remarks as appear to
me to be necessary, on the most important parts of
Mr. Mill's criticism of my *Philosophy*. I hope I have
avoided urging any thing in a contentious manner; as
I have certainly written with no desire of controversy,
but only with a view to offer to those who may be will-

ing to receive it, some explanation of portions of my previous writings. I have already said, that if this had not have been my especial object, I could with pleasure have noted the passages of Mr. Mill's *Logic* which I admire, rather than the points in which we differ. I will in a very few words refer to some of these points, as the most agreeable way of taking leave of the dispute.

I say then that Mr. Mill appears to me especially instructive in his discussion of the nature of the proof which is conveyed by the syllogism ; and that his doctrine, that the force of the syllogism consists in an *inductive assertion, with an interpretation added to it*, solves very happily the difficulties which baffle the other theories of this subject. I think that this doctrine of his is made still more instructive, by his excepting from it the cases of Scriptural Theology and of Positive Law (i. 260), as cases in which general propositions, not particular facts, are our original data. I consider also that the recognition of *Kinds* (i. 166) as classes in which we have, not a finite but an *inexhaustible* body of resemblances among individuals, and as groups made by nature, not by mere definition, is very valuable, as stopping the inroad to an endless train of false philosophy. I conceive that he takes the right ground in his answer to Hume's argument against miracles (ii. 183) : and I admire the acuteness with which he has criticized Laplace's tenets on the Doctrine of Chances, and the candour with which he has, in the second edition, acknowledged oversights on this subject made in the first. I think that much, I may almost say all, which he says on the subject of Language, is very philosophical ; for instance, what he says (ii. 238) of the way in which words acquire their meaning in common use. I especially admire the acuteness and force with which he has shown (ii. 255) how moral principles expressed in words degenerate into formulas, and yet how the formula cannot be rejected without a moral loss. This "perpetual oscillation in spiritual truths," as he happily terms it, has never, I think, been noted in the same broad manner, and

is a subject of most instructive contemplation. And
though I have myself refrained from associating moral
and political with physical science in my study of the
subject, I see a great deal which is full of promise
for the future progress of moral and political knowledge in Mr. Mill's sixth Book, "On the Logic of the
Moral and Political Sciences." Even his arrangement
of the various methods which have been or may be
followed in "the Social Science,"—"the Chemical or
Experimental Method," "the Geometrical or Abstract
Method," "the Physical or Concrete Deductive Method," "the Inverse Deductive or Historical Method,"
though in some degree fanciful and forced, abounds
with valuable suggestions; and his estimate of "the
interesting philosophy of the Bentham school," the
main example of "the geometrical method," is interesting and philosophical. On some future occasion,
I may, perhaps, venture into the region of which Mr.
Mill has thus essayed to map the highways: for it
is from no despair either of the great progress to be
made in such truth as that here referred to, or of
the effect of philosophical method in arriving at such
truth, that I have, in what I have now written, confined myself to the less captivating but more definite
part of the subject.

Of the Plurality of Worlds:

An Essay, also a Dialogue of the Same Subject

CHAPTER VI.

THE ARGUMENT FROM GEOLOGY.

1. I HAVE endeavored to explain that, according to the discoveries of geologists, the masses of which the surface of the earth is composed, exhibit indisputable evidence that, at different successive periods, the land and the waters which occupy it, have been inhabited by successive races of plants and animals ; which, when taken in large groups, according to the ascending or descending order of the strata, consist of species different from those above and below them. Many of these groups of species are of forms so different from any living things which now exist, as to give to the life of those ancient periods an aspect strangely diverse from that which life now displays, and to transfer us, in thought, to a creation remote in its predominant forms from that among which we live. I have shown also, that the life and successive generations of these groups of species, and the events by which the rocks which contain these remains have been brought into their present situation and condition, must have occupied immense intervals of time ;—intervals so large that they deserve to be compared, in their numerical expression, with the intervals of space which separate the planets and stars from each other. It has been seen, also, that the best geologists and natural his-

torians have not been able to devise any hypothesis to account
for the successive introduction of these new species into the
earth's population ; except the exercise of a series of acts of
creation, by which they have been brought into being; either
in groups at once, or in a perpetual succession of one or a few
species, which the course of long intervals of time might ac-
cumulate into groups of species. It is true, that some speculators have held that by the agency of natural causes, such as
operate upon organic forms, one species might be transmuted
into another; external conditions of climate, food, and the
like, being supposed to conspire with internal impulses and
tendencies, so as to produce this effect. This supposition is,
however, on a more exact examination of the laws of animal
life, found to be destitute of proof; and the doctrine of the
successive creation of species remains firmly established among
geologists. That the *extinction* of species, and of groups of
species, may be accounted for by natural causes, is a proposition much more plausible, and to a certain extent, probable ;
for we have good reason to believe that, even within the time
of human history, some few species have ceased to exist upon
the earth. But whether the extinction of such vast groups
of species as the ancient strata present to our notice, can be
accounted for in this way, at least without assuming the occurrence of great catastrophes, which must for a time, have
destroyed all forms of life in the district in which they occurred,
appears to be more doubtful. The decision of these questions,
however, is not essential to our purpose. What is important
is, that immense numbers of tribes of animals have tenanted
the earth for countless ages, before the present state of things
began to be.

2. The present state of things is that to which the existence
and the history of MAN belong ; and the remark which I now

have to make is, that the existence and the history of Man are facts of an entirely different order from any which existed in any of the previous states of the earth ; and that this history has occupied a series of years which, compared with geological periods, may be regarded as very brief and limited.

3. The remains of man are nowhere found in the strata which contain the records of former states of the earth. Skeletons of vast varieties of creatures have been disinterred from their rocky tombs ; but these cemeteries of nature supply no portion of a human skeleton. In earlier periods of natural science, when comparative anatomy was as yet very imperfectly understood, no doubt, many fossil bones were supposed to be human bones. The remains of giants and of antediluvians were frequent in museums. But a further knowledge of anatomy has made it appear that such bones all belong to animals, of one kind or another; often, to animals utterly different, in their form and skeleton, from man. Also some bones, really human, have been found petrified in situations in which petrification has gone on in recent times, and is still going on. Human skeletons, imbedded in rocks by this process, have been found in the island of Guadaloupe, and elsewhere. But this phenomenon is easily distinguishable from the petrified bones of other animals, which are found in rocks belonging to really geological periods ; and does not at all obliterate the distinction between the geological and the historical periods.

4. Indeed not bones only, but objects of art, produced by human workmanship, are found fossilized and petrified by the like processes ; and these, of course, belong to the historical period. Human bones, and human works, are found in such deposits as morasses, sand-banks, lava-streams, mounds of volcanic ashes ; and many of them may be of unknown, and,

compared with the duration of a few generations, of very
great antiquity ; but such deposits are distinguishable, generally without difficulty, from the strata in which the geologist
reads the records of former creations. It has been truly said,
that the geologist is an *Antiquary ;* for, like the antiquary, he
traces a past condition of things in the remains and effects of
it which still subsist; but it has also been truly said, at the
same time, that he is an antiquary *of a new Order ;* for the
remains which he studies are those which illustrate the history
of the earth, not of man. The geologist's antiquity is not that
of ornaments and arms, utensils and habiliments, walls and
mounds ; but of species and of genera, of seas and of mountains. It is true, that the geologist may have to study the
works of man, in order to trace the effects of causes which
produce the results which he investigates ; as when he ex-
amines the pholad-pierced pillars of Pateoli, to prove the rise
and the fall of the ground on which they stand ; or notes the
anchoring-rings in the wall of some Roman edifice, once a
maritime fort, but now a ruin remote from the sea ; or when
he remarks the streets in the towns of Scania, which are now
below the level of the Baltic,* and therefore show that the land
has sunk since these pavements were laid. But in studying
such objects, the geologist considers the hand of man as only
one among many agencies. Man is to him only one of the
natural causes of change.

5. And if, with the illustrious author to whom we have just
referred,† we liken the fossil remains, by which the geologist
determines the age of his strata, to the Medals and Coins in
which the antiquary finds the record of reigns and dynasties ;
we must still recollect that a *Coin* really discloses a vast body
of characteristics of man, to which there is nothing approach-

* Lyell, II. 420. [6th Ed.] † Cuvier.

ing in the previous condition of the world. For how much does a Coin or Medal indicate ? Property ; exchange ; government ; a standard of value ; the arts of mining, assaying, coining, drawing, and sculpture ; language, writing, and reckoning ; historical recollections, and the wish to be remembered by future ages. All this is involved in that small human .work, a Coin. If the fossil remains of animals may (as has been said) be termed Medals struck by Nature to record the epochs of her history ; Medals must be said to be, not merely, like fossil remains, records of material things ; they are the records of thought, purpose, society, long continued, long improved, supplied with multiplied aids and helps ; they are the permanent results, in a minute compass, of a vast progress, extending through all the ramifications of human life.

6. Not a coin merely, but any, the rudest work of human art, carries us far beyond the domain of mere animal life. There is no transition from man to animals. No doubt, there are races of men very degraded, barbarous, and brutish. No doubt there are kinds of animals which are very intelligent and sagacious ; and some which are exceedingly disposed to and adapted to companionship with man. But by elevating the intelligence of the brute, we do not make it become the intelligence of the man. By making man barbarous, we do not make him cease to be a man. Animals have their especial capacities, which may be carried very far, and may approach near to human sagacity, or may even go beyond it ; but the capacity of man is of a different kind. It is a capacity, not for becoming sagacious, but for becoming rational ; or rather it is a capacity which he has in virtue of being rational. It is a capacity of progress. In animals, however sagacious, however well trained, the progress in skill and knowledge is limited, and very narrowly limited. The creature soon reaches a boundary, beyond

which it cannot pass; and even if the acquired habits be transmitted by descent to another generation, (which happens in the
case of dogs and several other animals,) still the race soon
comes to a stand in its accomplishments. But in man, the possible progress from generation to generation, in intelligence
and knowledge, and we may also say, in power, is indefinite;
or if this be doubted, it is at least so vast, that compared with
animals, his capacity is infinite. And this capacity extends to
all races of men its characterizing efficacy : for we have good
reason to believe that there is no race of human beings who
may not, by a due course of culture, continued through generations, be brought into a community of intelligence and power
with the most intelligent and the most powerful races. This
seems to be well established, for instance, with regard to the
African negroes; so long regarded by most, by some probably regarded still, as a race inferior to Europeans. It has been
found that they are abundantly capable of taking a share in the
arts, literature, morality and religion of European peoples.
And we cannot doubt that, in the same manner, the native
Australians, or the Bushmen of the Cape of Good Hope, have
human faculties and human capacities; however difficult it
might be to unfold these, in one or two generations, into a
form of intelligence and civilization in any considerable degree
resembling our own.

7. It is not requisite for us, and it might lead to unnecessary
difficulties, to fix upon any one attribute of man, as peculiarly
characteristic, and distinguishing him from brutes. Yet it
would not be too much to say that man is, in truth, universally and specifically characterized by the possession of *Language*. It will not be questioned that language, in its highest
forms, is a wonderful vehicle and a striking evidence of the intelligence of man. His bodily organs can, by a few scarcely

perceptible motions, shape the air into sounds which express the kinds, properties, actions and relations of things, under thousands of aspects, in forms infinitely more general and recondite than those in which they present themselves to his senses;—and he can, by means of these forms, aided by the use of his senses, explore the boundless regions of space, the far recesses of past time, the order of nature, the working of the Author of nature. This man does, by the exercise of his Reason, and by the use of Language, a necessary implement of his Reason for such purposes.

8. That language, in such a stage, is a special character of man, will not be doubted. But it may be thought, there is little resemblance between Language in this exalted degree of perfection, and the seemingly senseless gibberish of the most barbarous tribes. Such an opinion, however, might easily be carried too far. All human language has in it the elements of indefinite intellectual activity, and the germs of indefinite development. Even the rudest kind of speech, used by savages, denotes objects by their kinds, their attributes, their relations, with a degree of generality derived from the intellect, not from the senses. The generality may be very limited; the relations which the human intellect is capable of apprehending may be imperfectly conveyed. But to denote kinds and attributes and actions and relations *at all*, is a beginning of generalization and abstraction ;—or rather, is far more than a beginning. It is the work of a faculty which can generalize and abstract ; and these mental processes once begun, the field of progress which is open to them is indefinite. Undoubtedly it may happen that weak and barbarous tribes are, for many generations, so hard pressed by circumstances, and their faculties so entirely absorbed in providing for the bare wants of the poorest life, that their thoughts may never travel to anything

beyond these, and their language may not be extended so as to be applicable to any other purposes. But this is not the standard condition of mankind. It is not, by such cases, that man, or that human nature, is to be judged. The normal condition of man is one of an advance beyond the mere means of subsistence, to the arts of life, and the exercise of thought in a general form. To some extent, such an advance has taken place in almost every region of the earth and in every age.

9. Perhaps we may often have a tendency to think more meanly than they deserve, of so-called barbarous tribes, and of those whose intellectual habits differ much from our own. We may be prone to regard ourselves as standing at the summit of civilization; and all other nations and ages, as not only occupying inferior positions, but positions on a slope which descends till it sinks into the nature of brutes. And yet how little does an examination of the history of mankind justify this view! The different stages of civilization, and of intellectual culture, which have prevailed among them, have had no appearance of belonging to one single series, in which the cases differed only as higher or lower. On the contrary, there have been many very different kinds of civilization, accompanied by different forms of art and of thought; showing how universally the human mind tends to such habits, and how rich it is in the modes of manifesting its innate powers. How different have been the forms of civilization among the Chinese, the Indians, the Egyptians, the Babylonians, the Mexicans, the Peruvians! Yet in all, how much was displayed of sagacity and skill, of perseverance and progress, of mental activity and grasp, of thoughtfulness and power. Are we, in thinking of these manifestations of human capacity, to think of them as only a stage between us and brutes? or are we to think so, even of the stoical Red Indians of North America, or the en-

5*

ergetic New Zealanders, and Caffres ?   And if not, why of the
African Negroes, or the Australians, or the Bushmen ?   We
may call their Language a jargon.   Very probable it would,
in its present form, be unable to express a great deal of what
we are in the habit of putting into language.   But can we refuse to believe that, with regard to matters with which they
are familiar, and on occasions where they are interested, they
would be to each other intelligible and clear ?   And if we suppose cases in which their affections and emotions are strongly
excited, (and affections and emotions at least we cannot deny
them,) can we not believe that they would be eloquent and impressive ?   Do we not know, in fact, that almost all nations
which we call savage, are, on such occasions, eloquent in their
own language ?   And since this is so, must not their language,
after all, be a wonderful instrument as well as ours ?   Since
it can convey one man's thoughts and emotions to many,
clothed in the form which they assume in his mind ; giving to
things, it may be, an aspect quite different from that which
they would have if presented to their own senses ; guiding
their conviction, warming their hearts, impelling their purposes ;—can language, even in such cases, be otherwise than a
wonderful produce of man's internal, of his mental, that is, of
his peculiarly *human* faculties ?   And is not language, therefore, even in what we regard as its lowest forms, an endow-
ment which completely separates man from animals which
have no such faculty ?—which cannot regard, or which cannot
convey, the impressions of the individual in any such general
and abstract form ?   Probably we should find, as those who
have studied the language of savages always have found, that
every such language contains a number of curious and subtle
practices,—*contrivances*, we cannot help calling them,—for
marking the relations, bearings and connections of words ; con-

trivances quite different from those of the languages which we think of as more perfect; but yet, in the mouths of those who use such speech, answering their purpose with great precision. But without going into such details, the use of any *articulate* language is, as the oldest Greeks spoke of it, a special and complete distinction of man as man.

10. It would be an obscure and useless labor, to speculate upon the question whether animals have among themselves anything which can properly be called *Language*. That they have anything which can be termed Language, in the sense in which we here speak of it, as admitting of general expressions, abstractions, address to numbers, eloquence, is utterly at variance with any interpretation which we can put upon their proceedings. The broad distinction of Instinct and Reason, however obscure it may be, yet seems to be most simply described, by saying, that animals do not apprehend their impressions under general forms, and that man does. Resemblance, and consequent association of impressions, may often show like generalization; but yet it is different. There is, in man's mind, a germ of general thoughts, suggested by resemblances, which is evolved and fixed in language; and by the aid of such an addition to the impressions of sense, man has thousands of intellectual pathways from object to object, from effect to cause, from fact to inference. His impressions are projected on a sphere of thought of which the radii can be prolonged into the farthest regions of the universe. Animals, on the contrary, are shut up in their sphere of sensation,— passing from one impression to another by various associations, established by circumstances; but still, having access to no wider intellectual region, through which lie lines of transition purely abstract and mental. That they have their modes of communicating their impressions and associations,

their affections and emotions, we know; but these modes of communication do not make a language; nor do they disturb the assignment of Language as a special character of man; nor the belief that man differs in his Kind, and we may say, using a larger phrase, in his Order, from all other creatures.

11. We may sometimes be led to assign much of the development of man's peculiar powers, to the influence of external circumstances. And that the development of those powers is so influenced, we cannot doubt; but their development only, not their existence. We have already said that savages, living a precarious and miserable life, occupied incessantly with providing for their mere bodily wants, are not likely to possess language, or any other characteristic of humanity, in any but a stunted and imperfect form. But, that manhood is debased and degraded under such adverse conditions, does not make man cease to be man. Even from such an abject race, if a child be taken and brought up among the comforts and means of development which civilized life supplies, he does not fail to show that he possesses, perhaps in an eminent degree, the powers which specially belong to man. The evidences of human tendencies, human thoughts, human capacities, human affections and sympathies, appear conspicuously, in cases in which there has been no time for external circumstances to operate in any great degree, so as to unfold any difference between the man and the brute; or in which the influence of the most general of external agencies, the impressions of several of the senses, have been intercepted. Who that sees a lively child, looking with eager and curious eyes at every object, uttering cries that express every variety of elementary human emotion in the most vivacious manner, exchanging looks and gestures, and inarticulate sounds, with his nurse, can doubt that already he possesses the germs of

human feeling, thought and knowledge? that already, before he can form or understand a single articulate word, he has within him the materials of an infinite exuberance of utterance, and an impulse to find the language into which such utterance is to be moulded by the law of his human nature? And perhaps it may have happened to others, as it has to me, to know a child who had been both deaf, dumb, and blind, from a very early age. Yet she, as years went on, disclosed a perpetually growing sympathy with the other children of the family in all their actions, with which of course she could only acquaint herself by the sense of touch. She sat, dressed, walked, as they did; even imitated them in holding a book in her hand when they read, and in kneeling when they prayed. No one could look at the change which came over her sightless countenance, when a known hand touched hers, and doubt that there was a human soul within the frame. The human soul seemed not only to be there, but to have been fully developed; though the means by which it could receive such communications as generally constitute human education, were thus cut off. And such modes of communication with her companions as had been taught her, or as she had herself invented, well bore out the belief, that her mind was the constant dwelling-place, not only of human affections, but of human thoughts. So plainly does it appear that human thought is not produced or occasioned by external circumstances only; but has a special and indestructible germ in human nature.

12. I have been endeavoring to illustrate the doctrine that man's nature is different from the nature of other animals; as subsidiary to the doctrine that the Human Epoch of the earth's history is different from all the preceding Epochs. But in truth, this subsidiary proposition is not by any means neces-

sary to my main purpose. Even if barbarous and savage
tribes, even if men under unfavorable circumstances, be little
better than the brutes, still no one will doubt that the most
civilized races of mankind, that man under the most favorable
circumstances, is far, is, indeed, immeasurably elevated above
the brutes. The history of man includes not only the history
of Scythians and Barbarians, Australians and Negroes, but of
ancient Greeks and of modern Europeans; and therefore there
can be no doubt that the period of the Earth's history, which
includes the history of man, is very different indeed from any
period which preceded that. To illustrate the peculiarity, the
elevation, the dignity, the wonderful endowments of man, we
might refer to the achievements, the recorded thoughts and
actions, of the most eminent among those nations;—to their
arts, their poetry, their eloquence; their philosophers, their
mathematicians, their astronomers; to the acts of virtue and
devotion, of patriotism, generosity, obedience, truthfulness,
love, which took place among them;—to their piety, their
reverence for the deity, their resignation to his will, their hope
of immortality. Such characteristic traits of man as man,
(which all examples of intelligence, virtue, and religion, are,)
might serve to show that man is, in a sense quite different
from other creatures, " fearfully and wonderfully made;" but
I need not go into such details. It is sufficient for my purpose to sum up the result in the expressions which I have
already used; that man is an intellectual, moral, religious,
and spiritual being.

13. But the existence of man upon the earth being thus an
event of an order quite different from any previous part of the
earth's history, the question occurs, how long has this state of
things endured? What period has elapsed since this creature,
with these high powers and faculties, was placed upon the

earth? How far must we go backward in time, to find the beginning of his wonderful history?—so utterly wonderful compared with anything which had previously occurred. For as to that point, we cannot feel any doubt. The wildest imagination cannot suggest that corals and madrepores, oysters and sepias, fishes and lizards, may have been rational and moral creatures; nor even those creatures which come nearer to human organization; megatheriums and mastodons, extinct deer and elephants. Undoubtedly the earth, till the existence of man, was a world of mere brute creatures. How long then has it been otherwise? How long has it been the habitation of a rational, reflective, progressive race? Can we by any evidence, geological or other, approximate to the beginning of the Human History?

14. This is a large and curious question, and one on which a precise answer may not be within our reach. But an answer not precise, an approximation, as we have suggested, may suffice for our purpose. If we can determine, in some measure, the order and scale of the period during which man has occupied the earth, the determination may serve to support the analogy which we wish to establish.

15. The geological evidence with regard to the existence of man is altogether negative. Previous to the deposits and changes which we can trace as belonging obviously to the present state of the earth's surface, and the operation of causes now existing, there is no vestige of the existence of man, or of his works. As was long ago observed,* we do not find, among the shells and bones which are so abundant in the older strata, any weapons, medals, implements, structures, which speak to us of the hand of man, the workman. If we look forwards ten or twenty thousand years, and suppose the existing works

* By Bishop Berkeley. See Lyell, III. 346.

of man to have been, by that time, ruined and covered up by masses of rubbish, inundations, morasses, lava-streams, earthquakes; still, when the future inhabitant of the earth digs into and explores these coverings, he will discover innumerable monuments that man existed so long ago. The materials of many of his works, and the traces of his own mind, which he stamps upon them, are as indestructible as the shells and bones which give language to the oldest work. Indeed, in many cases the oldest fossil remains are the results of objects of seemingly the most frail and perishable material;—of the most delicate and tender animal and vegetable tissues and filaments. That no such remains of textures and forms, moulded by the hand of man, are anywhere found among these, must be accepted as indisputable evidence that man did not exist, so as to be contemporary with the plants and animals thus commemorated. According to geological evidence, the race of man is a novelty upon the earth;—something which has succeeded to all the great geological changes.

16. And in this, almost all geologists are agreed. Even those who hold that, in other ways, the course of change has been uniform;—that even the introduction of man, as a new species of animal, is only an event of the same kind as myriads of like events which have occurred in the history of the earth;—still allow that the introduction of man, as a moral being, is an event entirely different from any which had taken place before; and that event is, geologically speaking, recent. The changes of which we have spoken, as studied by the geologist in connection with the works of man, the destruction of buildings on sea-coasts by the incursions of the ocean, the removal of the shore many miles away from ancient harbors, the overwhelming of cities by earthquakes or volcanic eruptions; however great when compared with the changes which

371

take place in one or two generations ; are minute and infinitesimal, when put in comparison with the changes by which ranges of mountains and continents have been brought into being, one after another, each of them filled with the remains of different organic creations.

17. Further than this, geology does not go on this question. She has no chronometer which can tell us when the first buildings were erected, when man first dwelt in cities, first used implements or arms; still less, language and reflection. Geology is compelled to give over the question to History. The external evidences of the antiquity of the species fail us, and we must have recourse to the internal. Nature can tell us so little of the age of man, that we must inquire what he can tell us himself.

42. That men are, only in a comparatively small number of cases, intellectual, moral, religious, and spiritual, in the degree which I have described, does not, by any means, deprive our argument of its force. The capacity of man is, that he may become this ; and such a capacity may well make him a special object in the eyes of Him under whose guidance and by whose aid, such a development and elevation of his nature is open to him. However imperfect and degraded, however unintellectual, immoral, irreligious, and unspiritual, a great part of mankind may be, still they all have the germs of such an elevation of their nature; and a large portion of them make, we cannot doubt, no small progress in this career of advancement to a spiritual condition. And with such capaci-

ties, and such practical exercise of those capacities, we can have no difficulty in believing, if the evidence directs us to believe, that that part of the creation in which man has his present appointed place, is the special field of God's care and love ; by whatever wastes of space, and multitudes of material bodies, it may be surrounded ; by whatever races it may have been previously occupied, of brutes that perish, and that, compared with man, can hardly be said to have lived.

CHAPTER XII.

THE UNITY OF THE WORLD.

1. The two doctrines which we have here to weigh against each other are the Plurality of Worlds, and the Unity of the World. In so saying, we include in our present view, a necessary part of the conception of a *World,* a collection of intelligent creatures : for even if the suppositions to which we have been led, respecting the kind of unintelligent living things which may inhabit other parts of the Universe, be conceived to be probable ; such a belief will have little interest for most persons, compared with the belief of other worlds, where reside intelligence, perception of truth, recognition of moral Law, and reverence for a Divine Creator and Governor. In looking outwards at the Universe, there are certain aspects which suggest to man, at first sight, a conjecture that there may be other bodies like the Earth, tenanted by other creatures like man. This conjecture, however, receives no confirmation from a closer inquiry, with increased means of observation. Let us now look inwards, at the constitution of man ; and consider some characters of his nature, which seem to remove or lessen the difficulties which we may at first feel, in regarding the Earth as, in a unique and special manner, the field of God's Providence and Government.

2. In the first place, the Earth, as the abode of man, the in-

tellectual creature, contains a being, whose mind is, in some measure, of the same nature as the Divine Mind of the Creator. The Laws which man discovers in the Creation must be Laws known to God. The truths,—for instance the truths of geometry,—which man sees to be true, God also must see to be true. That there were, from the beginning, in the Creative Mind, Creative Thoughts, is a doctrine involved in every intelligent view of Creation.

3. This doctrine was presented by the ancients in various forms ; and the most recent scientific discoveries have supplied new illustrations of it. The mode in which Plato expressed the doctrine which we are here urging was, that there were in the Divine Mind, before or during the work of creation, certain archetypal Ideas, certain exemplars or patterns of the world and its parts, according to which the work was performed : so that these Ideas or Exemplars existed in the objects around us being in so many cases discernible by man, and being the proper objects of human reason. If a mere metaphysician were to attempt to revive this mode of expressing the doctrine, probably his speculations would be disregarded, or treated as a pedantic resuscitation of obsolete Platonic dreams. But the adoption of such language must needs be received in a very different manner, when it proceeds from a great discoverer in the field of natural knowledge : when it is, as it were, forced upon *him*, as the obvious and appropriate expression of the result of the most profound and comprehensive researches into the frame of the whole animal creation. The recent works of Mr. Owen, and especially one work, *On the Nature of Limbs*, are full of the most energetic and striking passages, inculcating the doctrine which we have been endeavoring to maintain. We may take the liberty of enriching our pages with one passage bearing upon the present part of the subject.

"If the world were made by any antecedent Mind or Understanding, that is by a Deity, then there must needs be an Idea and Exemplar of the whole world before it was made, and consequently actual knowledge, both in the order of Time and Nature, before Things. But conceiving of knowledge as it was got by their own finite minds, and ignorant of any evidence of an ideal Archetype for the world or any part of it, they [the Democritic Philosophers who denied a Divine Creative Mind] affirmed that there was none, and concluded that there could be no knowledge or mind before the world was, as its cause." Plato's assertion of Archetypal Ideas was a protest against this doctrine, but was rather a guess, suggested by the nature of mathematical demonstration, than a doctrine derived from a contemplation of the external world.

"Now however," Mr. Owen continues, "the recognition of an ideal exemplar for the vertebrated animals proves that the knowledge of such a being as Man must have existed before Man appeared. For the Divine Mind which planned the Archetypal also foreknew all its modifications. The Archetypal Idea was manifested in the flesh under divers modifications upon this planet, long prior to the existence of those animal species which actually exemplify it. To what natural or secondary causes the orderly succession and progression of such organic phenomena may have been committed, we are as yet ignorant. But if without derogation to the Divine Power, we may conceive such ministers and personify them by the term *Nature*, we learn from the past history of our globe that she has advanced with slow and stately steps, guided by the archetypal light amidst the wreck of worlds, from the first embodiment of the vertebrate idea, under its old ichthyic vestment, until it became arrayed in the glorious garb of the human form."

4. Law implies a Lawgiver, even when we do not see the object of the Law; even as Design implies a Designer, when we do not see the object of the Design. The Laws of Nature are the indications of the operation of the Divine Mind; and are revealed to us, as such, by the operations of our minds, by which we come to discover them. They are the utterances of the Creator, delivered in language which we can understand; and being thus Language, they are the utterances of an Intelligent Spirit.

5. It may seem to some persons too bold a view, to identify, so far as we thus do, certain truths as seen by man, and as seen by God :*—to make the Divine Mind thus cognizant of the truths of geometry, for instance. If any one has such a scruple, we may remark that truth, when of so luminous and stable a kind as are the truths of geometry, must be alike *Truth* for all minds, even for the highest. The mode of arriving at the knowledge of such truths, may be very different, even for different human minds;—deduction for some;—intuition for others. But the intuitive apprehension of necessary truth is an act so purely intellectual, that even in the Supreme Intellect, we may suppose that it has its place. Can we conceive otherwise, than that God does contemplate the universe as existing in space, since it really does so;—and subject to the relations of space, since these are as real as space itself? We are well aware that the Supreme Being must contemplate the world under many other aspects than this;—even man does so. But that does not prevent the truths, which belong to the aspect of the world, contemplated as existing in space, from being truths, regarded as such, even by the Divine Mind.

* Among the most recent expositors of this doctrine we may place M. Henri Martin, whose *Philosophie Spiritualiste de la Nature* is full of striking views of the universe in its relation to God. (Paris. 1849.)

6. If these reflections are well founded, as we trust they will, on consideration, be seen to be, we may adopt many of the expressions by which philosophers heretofore have attempted to convey similar views ; for in fact, this view, in its general bearing at least, is by no means new. The Mind of Man is a partaker of the thoughts of the Divine Mind. The Intellect of Man is a spark of the Light by which the world was created. The Ideas according to which man builds up his knowledge, are emanations of the archetypal Ideas according to which the work of creation was planned and executed. These, and many the like expressions, have been often used ; and we now see, we may trust, that there is a great philosophical truth, which they all tend to convey ; and this truth shows at the same time, how man may have some knowledge respecting the Laws of Nature, and how this knowledge may, in some cases, seem to be a knowledge of necessary relations, as in the case of space.*

* Most readers who have given any attention to speculations of this kind, will recollect Newton's remarkable expressions concerning the Deity : "Æternus est et infinitus, omnipotens et omnisciens ; id est, durat ab æterno in æternum, et adest ab infinito in infinitum . . . Non est æternitas et infinitas, sed æternus et infinitus ; non est duratio et spatium, sed durat et adest. Durat semper et adest ubique, et existendo semper et ubique durationem et spatium constituit."

To say that God by existing always and everywhere *constitutes duration and space*, appears to be a form of expression better avoided. Besides that it approaches too near to the opinion, which the writer rejects, that He *is* duration and space, it assumes a knowledge of the nature of the Divine existence, beyond our means of knowing, and therefore rashly. It appears to be safer, and more in conformity with what we really know, to say, not that the existence of God constitutes time and space ; but that God has constituted *man*, so that *he* can apprehend the works of creation, only as existing in time and space. That God has constituted time and space as conditions of man's knowledge of the creation, is certain : that God has constituted time and space as results of his own existence in any other way, *we* cannot know.

7. Now, the views to which we have been led, bear very strongly upon that argument. For if man, when he attains to a knowledge of such laws, is really admitted, in some degree, to the view with which the Creator himself beholds his creation;—if we can gather, from the conditions of such knowledge, that his intellect partakes of the Nature of the Supreme Intellect;—if his Mind, in its clearest and largest contemplations, harmonizes with the Divine Mind;—we have, in this, a reason which may well seem to us very powerful, why, even if the Earth alone be the habitation of intelligent beings, still, the great work of Creation is not wasted. If God have placed upon the earth a creature who can so far sympathize with Him, if we may venture upon the expression;—who can raise his intellect into some accordance with the Creative Intellect; and that, not once only, nor by few steps, but through an indefinite gradation of discoveries, more and more comprehensive, more and more profound; each, an advance, however slight, towards a Divine Insight;—then, so far as intellect alone (and we are here speaking of intellect alone) can make Man a worthy object of all the vast magnificence of Creative Power, we can hardly shrink from believing that he is so.

8. We may remark further, that this view of God, as the Author of the Laws of the Universe, leads to a view of all the phenomena and objects of the world, as the work of God; not a work made, and laid out of hand, but a field of his present activity and energy. And such a view cannot fail to give an aspect of dignity to all that is great in creation, and of beauty to all that is symmetrical, which otherwise they could not have. Accordingly, it is by calling to their thoughts the presence of God as suggested by scenes of grandeur or splendor, that poets often reach the sympathies of their readers. And this dignity and sublimity appear especially to belong to the larger

objects, which are destitute of conscious life ; as the mountain, the glacier, the pine-forest, the ocean ; since in these, we are, as it were, alone with God, and the only present witnesses of His mysterious working.

9. Now if this reflection be true, the vast bodies which hang in the sky, at such immense distances from us, and roll on their courses, and spin round their axles with such exceeding rapidity ; Jupiter and his array of Moons, Saturn with his still larger host of Satellites, and with his wonderful Ring, and the other large and distant Planets, will lose nothing of their majesty, in our eyes, by being uninhabited ; any more than the summer-clouds, which perhaps are formed of the same materials, lose their dignity from the same cause ;—any more than our Moon, one of the tribe of satellites, loses her soft and tender beauty, when we have ascertained that she is more barren of inhabitants than the top of Mount Blanc. However destitute the planets and moons and rings may be of inhabitants, they are at least vast scenes of God's presence, and of the activity with which he carries into effect, everywhere, the laws of nature. The light which comes to us from them is transmitted according to laws which He has established, by an energy which He maintains. The remotest planet is not devoid of life, for God lives there. At each stage which we make, from planet to planet, from star to star, into the regions of infinity, we may say, with the patriarch, " Surely God is here, and I knew it not." And when those who question the habitability of the remote planets and stars are reproached as presenting a view of the universe, which takes something from the magnificence hitherto ascribed to it, as the scene of God's glory, shown in the things which He has created ; they may reply, that they do not at all disturb that glory of the creation which arises from its being, not only the product, but the constant field of God's

activity and thought, wisdom and power ; and they may perhaps ask, in return, whether the dignity of the Moon would be
greatly augmented if her surface were ascertained to be abundantly peopled with lizards ; or whether Mount Blanc would be
more sublime, if millions of frogs were known to live in the
crevasses of its glaciers.

10. Again : the Earth is a scene of Moral Trial. Man is
subject to a Moral Law ; and this Moral Law is a Law of
which God is the Legislator. It is a law which man has the
power of discovering, by the use of the faculties which God has
given him. By considering the nature and consequences of
actions, man is able to discern, in a great measure, what is right
and what is wrong ;—what he ought and what he ought not to
do ;—what his duty and virtue, what his crime and vice. Man
has a Law on such subjects, written on his heart, as the Apostle
Paul says. He has a conscience which accuses or excuses
him ; and thus, recognizes his acts as worthy of condemnation
or approval. And thus, man is, and knows himself to be, the
subject of Divine Law, commanding and prohibiting ; and is
here, in a state of probation, as to how far he will obey or
disobey this Law. He has impulses, springs of action, which
urge him to the violation of this Law. Appetite, Desire,
Anger, Lust, Greediness, Envy, Malice, impel him to courses
which are vicious. But these impulses he is capable of resist-
ing and controlling ;—of avoiding the vices and practising the
opposite virtues ;—and of rising from one stage of Virtue to
another, by a gradual and successive purification and elevation
of the desires, affections and habits, in a degree, so far as we
know, without limit.

16. Perhaps it may be said, that all which we have urged to show that other animals, in comparison with man, are less worthy objects of creative design, may be used as an argument to prove that other planets are tenanted by men, or by moral and intellectual creatures like man ; since, if the creation of *one* world of such creatures exalts so highly our views of the dignity and importance of the plan of creation, the belief in *many* such worlds must elevate still more our sentiments of admiration and reverence of the greatness and goodness of the Creator ; and must be a belief, on that account, to be accepted and cherished by pious minds.

17. To this we reply, that we cannot think ourselves authorized to assert cosmological doctrines, selected arbitrarily by ourselves, on the ground of their exalting our sentiments of admiration and reverence for the Deity, *when the weight of all the evidence which we can obtain respecting the constitution of the universe is against them.* It appears to us, that to discern one great scheme of moral and religious government, which is the spiritual centre of the universe, may well suffice for the religious sentiments of men in the present age ; as in former ages such a view of creation was sufficient to overwhelm men with feelings of awe, and gratitude, and love ; and to make them confess, in the most emphatic language, that all such feelings were an inadequate response to the view of the scheme of Providence which was revealed to them. The thousands of millions of inhabitants of the Earth to whom the effects of the Divine Plan extend, will not seem, to the greater part of religious persons, to need the addition of more, to fill our minds with sufficiently vast and affecting contemplations, so far as we are capable of pursuing such contemplations. The possible extension of God's spiritual kingdom upon the earth will probably appear to them a far more inter-

esting field of devout meditation, than the possible addition to it of the inhabitants of distant stars, connected in some inscrutable manner with the Divine Plan.

18. To justify our saying that the weight of the evidence is against such cosmological doctrines, we must recall to the reader's recollection the whole course of the argument which we have been pursuing.

It is a possible conjecture, at first, that there may be other Worlds, having, as this has, their moral and intellectual attributes, and their relations to the Creator. It is also a possible conjecture, that this World, having such attributes, and such relations, may, on that account, be necessarily unique and incapable of repetition, peculiar, and spiritually central. These two opposite possibilities may be placed, at first, front to front, as balancing each other. We must then weigh such evidence and such analogies as we can find on the one side or on the other. We see much in the intellectual and moral nature of man, and in his history, to confirm the opinion that the human race is thus unique, peculiar and central. In the views which Religion presents, we find much more, tending the same way, and involving the opposite supposition in great difficulties. We find, in our knowledge of what we ourselves are, reasons to believe that if there be, in any other planet, intellectual and moral beings, they must not only be *like* men, but must *be* men, in all the attributes which we can conceive as belonging to such beings. And yet to suppose other groups of the human species, in other parts of the universe, must be allowed to be a very bold hypothesis, to be justified only by some positive evidence in its favor. When from these views, drawn from the attributes and relations of man, we turn to the evidence drawn from physical conditions, we find very strong reason to believe that, so far as the Solar System is concerned, the Earth *is*, with regard to the conditions of life, in a peculiar and central position ; so that the conditions of any life approaching at all to human life, exist on the Earth alone. As to other systems which may circle other suns, the possibility of their being inhabited by men, remains, as at first, a mere conjecture, without any trace of confirmatory evidence. It was suggested at first by the supposed analogy of other stars to our sun ; but this analogy has not been verified in any instance ; and has been, we conceive, shown in many cases, to vanish altogether. And that there may be such a plan of creation,—one in which the moral and intelligent race of man is the climax and central point to which innumerable races of mere unintelligent species tend,—we have the most striking evidence, in the history of our own earth, as disclosed by geology. We are left, therefore, with nothing to cling to, on one side, but the bare possibility that some of the stars are the centres of systems like the Solar System ;—an opinion founded upon the single fact, shown to be highly ambiguous, of those stars being self-luminous ; and to this possibility, we oppose all the considerations, flowing from moral, historical, and religious views, which represent the human race as unique and peculiar. The force of these considerations will, of course, be different in different minds, according to the importance which each person attaches to such moral, historical, and religious views ; but whatever the weight of them may be deemed, it is to be recollected that we have on the other side a bare possibility, a mere conjecture ; which, though suggested at first by astronomical discoveries, all more recent astronomical researches have failed to confirm in the smallest degree. In this state of our knowledge, and with such grounds of belief, to dwell upon the Plurality of Worlds of intellectual and moral creatures, as a highly probable doctrine, must, we think, be held to be eminently rash and unphilosophical.

XII. *Of the Transformation of Hypotheses in the History of Science.*
By W. WHEWELL, D.D., *Master of Trinity College.*

[Read *May* 19, 1851.]

1. The history of science suggests the reflection that it is very difficult for the same person at the same time to do justice to two conflicting theories. Take for example the Cartesian hypothesis of vortices and the Newtonian doctrine of universal gravitation. The adherents of the earlier opinion resisted the evidence of the Newtonian theory with a degree of obstinacy and captiousness which now appears to us quite marvellous: while on the other hand, since the complete triumph of the Newtonians, *they* have been unwilling to allow any merit at all to the doctrine of vortices. It cannot but seem strange, to a calm observer of such changes, that in a matter which depends upon mathematical proofs, the whole body of the mathematical world should pass over, as in this and similar cases they seem to have done, from an opinion confidently held, to its opposite. No doubt this must be, in part, ascribed to the lasting effects of education and early prejudice. The old opinion passes away with the old generation: the new theory grows to its full vigour when its congenital disciples grow to be masters. John Bernoulli continues a Cartesian to the last; Daniel, his son, is a Newtonian from the first. Newton's doctrines are adopted at once in England, for they are the solution of a problem at which his contemporaries have been labouring for years. They find no adherents in France, where Descartes is supposed to have already explained the constitution of the world; and Fontenelle, the secretary of the Academy of Sciences at Paris, dies a Cartesian seventy years after the publication of Newton's *Principia*. This is, no doubt, a part of the explanation of the pertinacity with which opinions are held, both before and after a scientific revolution: but this is not the whole, nor perhaps the most instructive aspect of the subject. There is another feature in the change, which explains, in some degree, how it is possible that, in subjects, mainly at least mathematical, and therefore claiming demonstrative evidence, mathematicians should hold different and even opposite opinions. And the object of the present paper is to point out this feature in the successions of theories, and to illustrate it by some prominent examples drawn from the history of science.

2. The feature to which I refer is this; that when a prevalent theory is found to be untenable, and consequently, is succeeded by a different, or even by an opposite one, the change is not made suddenly, or completed at once, at least in the minds of the most tenacious adherents of the earlier doctrine; but is effected by a transformation, or series of transformations, of the earlier hypothesis, by means of which it is gradually brought nearer and nearer to the second; and thus, the defenders of the ancient doctrine are able to go on as if still asserting their first opinions, and to continue to press their points of advantage, if they have any, against the new theory. They borrow, or imitate, and in some way accommodate to their original hypothesis, the new explanations which the new theory gives, of the observed facts; and thus they maintain a sort of verbal consistency; till the original hypothesis becomes inextricably confused, or breaks down under the weight of the auxiliary hypotheses thus fastened upon it, in order to make it consistent with the facts.

This often-occurring course of events might be illustrated from the history of the astronomical theory of epicycles and eccentrics, as is well known. But my present purpose is to give one or two brief illustrations of a somewhat similar tendency from other parts of scientific history; and in the first place, from that part which has already been referred to, the battle of the Cartesian and Newtonian systems.

3. The part of the Cartesian system of vortices which is most familiarly known to general readers is the explanation of the motions of the planets by supposing them carried round the sun by a kind of whirlpool of fluid matter in which they are immersed: and the explanation of the motions of the satellites round their primaries by similar subordinate whirlpools, turning round the primary, and carried, along with it, by the primary vortex. But it should be borne in mind that a part of the Cartesian hypothesis which was considered quite as important as the cosmical explanation, was the explanation which it was held to afford of terrestrial gravity. Terrestrial gravity was asserted to arise from the motion of the vortex of subtile matter which revolved round the earth's axis and filled the surrounding space. It was maintained that by the rotation of such a vortex, the particles of the subtile matter would exert a centrifugal force, and by virtue of that force, tend to recede from the center: and it was held that all bodies which were near the earth, and therefore immersed in the vortex, would be pressed towards the center by the effort of the subtile matter to recede from the center[*].

These two assumed effects of the Cartesian vortices—to carry bodies in their stream, as straws are carried round by a whirlpool, and to press bodies to the center by the centrifugal effort of the whirling matter—must be considered separately, because they were modified separately, as the progress of discussion drove the Cartesians from point to point. The former effect indeed, the *dragging* force of the vortex, as we may call it, would not bear working out on mechanical principles at all; for as soon as the law of motion was acknowledged (which Descartes himself was one of the loudest in proclaiming), that a body in motion keeps all the motion which it has, and receives in addition all that is impressed upon it;—as soon, in short, as philosophers rejected the notion of an inertness in matter which constantly retards its movements,—it was plain that a planet perpetually dragged onwards in its orbit by a fluid moving quicker than itself, must be perpetually accelerated; and therefore could not follow those constantly-recurring cycles of quicker and slower motion which the planets exhibit to us.

The Cartesian mathematicians, then, left untouched the calculation of the progressive motion of the planets; and, clinging to the assumption that a vortex would produce a tendency of bodies to the center, made various successive efforts to construct their vortices in such a manner that the centripetal forces produced by them should coincide with those which the phenomena required, and therefore of course, in the end, with those which the Newtonian theory asserted.

In truth, the Cartesian vortex was a bad piece of machinery for producing a central force: from the first, objections were made to the sufficiency of its mechanism, and most of these objections were very unsatisfactorily answered, even granting the additional machinery which its defenders demanded. One formidable objection was soon started, and continued to the last to be the torment of the Cartesians. If terrestrial gravity, it was urged, arise from the centrifugal force of a vortex which revolves about the earth's axis, terrestrial gravity ought to act in planes perpendicular to the earth's axis, instead of tending to the earth's center. This objection was taken by James Bernoulli[*], and by Huyghens[†] not long after the publication of Descartes's *Principia*. Huyghens (who adopted the theory of vortices with modifications of his

---

[*]Cartes. *Princip.* IV. 23   [†]*De la Cause de la Pesanteur* (1689), p. 135.

[*]Jac. Bernoulli, *Nouvelles Pensées sur le Système de M. Descartes*, Op. T. I. p. 239 (1686).

own) supposes that there are particles of the fluid matter which move about the earth in every possible direction, within the spherical space which includes terrestrial objects; and that the greater part of these motions being in spherical surfaces concentric with the earth, produces a tendency towards the earth's center.

This was a procedure tolerably arbitrary, but it was the best which could be done. Saurin, a little later‡, gave nearly the same solution of this difficulty. The solution, identifying a vortex of some kind with a central force, made the hypothesis of vortices applicable wherever central forces existed; but then, in return, it deprived the image of a vortex of all that clearness and simplicity which had been its first great recommendation.

But still there remained difficulties not less formidable. According to this explanation of gravity, since the tendency of bodies to the earth's center arose from the superior centrifugal force of the whirling matter which pushed them inward as water pushes a light body upward, bodies ought to tend more strongly to the center in proportion as they are less dense. The rarest bodies should be the heaviest; contrary to what we find.

Descartes's original solution of this difficulty has a certain degree of ingenuity. According to him *(Princip.* IV. 23) a terrestrial body consists of particles of the *third element,* and the more it has of such particles, the more it excludes the parts of the *celestial matter,* from the revolution of which matter gravity arises; and therefore the denser is the terrestrial body, and the heavier it will be.

But though this might satisfy him, it could not satisfy the mathematicians who followed him, and tried to reduce his system to calculation on mechanical principles. For how could they do this, if the celestial matter, by the operation of which the phenomena of force and motion were produced, was so entirely different from ordinary matter, which alone had supplied men with experimental illustrations of mechanical principles? In order that the celestial matter, by its whirling, might produce the gravity of heavy bodies, it was mechanically necessary that it must be very dense; and *dense* in the ordinary sense of the term; for it was by regarding density in the ordinary sense of the term that the mechanical necessity had been established.

The Cartesians tried to escape this result (Huyghens, *Pesanteur,* p. 161, and John Bernoulli, *Nouvelles Pensées,* Art. 31) by saying that there were two meanings of *density* and *rarity;* that some fluids might be rare by having their particles far asunder, others, by having their particles very small though in contact. But it is difficult to think that they could, as persons well acquainted with mechanical principles, satisfy themselves with this distinction; for they could hardly fail to see that the mechanical effect of any portion of fluid depends upon the total mass moved, not on the size of its particles.

Attempts made to exemplify the vortices experimentally only shewed more clearly the force of this difficulty. Huyghens had found that certain bodies immersed in a whirling fluid tended to the center of the vortex. But when Saulmon* a little later made similar experiments, he had the mortification of finding that the heaviest bodies had the greatest tendency to recede from the axis of the vortex. "The result is," as the Secretary of the Academy (Fontenelle) says, "exactly the opposite of what we could have wished, for the [Cartesian] system of gravity: but we are not to despair; sometimes in such researches disappointment leads to ultimate success."

---

‡Journal des Savans, 1703. Mém. Acad. Par. 1709.

*Acad. Par. 1714. *Hist.* p. 106.

Bullfinger, in 1726 (Acad. Petrop.), conceived that by making a sphere revolve at the same time about two axes at right angles to each other, every particle would describe a great circle; but this is not so.

But, passing by this difficulty, and assuming that in some way or other a centripetal force arises from the centrifugal force of the vortex, the Cartesian mathematicians were naturally led to calculate the circumstances of the vortex on mechanical principles; especially Huyghens, who had successfully studied the subject of centrifugal force. Accordingly, in his little treatise on the *Cause of Gravitation,* (p. 143) he calculates the velocity of the fluid matter of the vortex, and finds that, at a point in the equator, it is 17 times the velocity of the earth's rotation.

It may naturally be asked, how it comes to pass that a stream of fluid, dense enough to produce the gravity of bodies by its centrifugal force, moving with a velocity 17 times that of the earth (and therefore moving round the earth in 85 minutes), does not sweep all terrestrial objects before it. But to this Huyghens had already replied (p. 137), that there are particles of the fluid moving *in all directions,* and therefore that they neutralize each other's action, so far as lateral motion is concerned.

And thus, as early as this treatise of Huyghens, that is, in three years from the publication of Newton's *Principia,* a vortex is made to mean nothing more than some machinery or other for producing a central force. And this is so much the case, that Huyghens commends (p. 165), as confirming his own calculation of the velocity of his vortex, Newton's proof that at the Moon's orbit the centripetal force is equal to the centrifugal; and that thus, this force is less than the centripetal force at the earth's surface in the inverse proportion of the squares of the distances.

John Bernoulli, in the same manner, but with far less clearness and less candour, has treated the hypothesis of vortices as being principally a hypothetical cause of central force. He had repeated occasions given him of propounding his inventions for propping up the Cartesian doctrine, by the subjects proposed for prizes by the Paris Academy of Sciences; in which competition Cartesian speculations were favourably received. Thus the subject of the Prize Essays for 1730 was, the explanation of the Elliptical Form of the planetary orbits and of the Motion of their Aphelia, and the prize was assigned to John Bernoulli, who gave the explanation on Cartesian principles. He explains the elliptical figure, not as Descartes himself had done, by supposing the vortex which carries the planet round the sun to be itself squeezed into an elliptical form by the pressure of contiguous vortices; but he supposes the planet, while it is carried round by the vortex, to have a limited oscillatory motion to and from the center, produced by its being originally, not at the distance at which it would float in equilibrium in the vortex, but above or below that point. On this supposition, the planet would oscillate to and from the center, Bernoulli says, like the mercury when deranged in a barometer: and it is evident that such an oscillation, combined with a motion round the center, might produce an oval curve, either with a fixed or with a moveable aphelion. All this however merely amounts to a possibility that the oval *may* be an ellipse, not to a proof that it will be so; nor does Bernoulli advance further.

It was necessary that the vortices should be adjusted in such a manner as to account for Kepler's laws; and this was to be done by making the velocity of each stratum of the vortex depend in a suitable manner on its radius. The Abbé de Molières attempted this on the supposition of elliptical vortices, but could not reconcile Kepler's first two laws, of equal elliptical areas in equal times, with his third law, that the squares of the periodic times are as the cubes of the mean distances*. Bernoulli, with his circular vortices, could accommodate the velocities at different distances so that they should explain Kepler's laws. He pretended to prove that Newton's

---

* Acad. Par. 1733.

investigations respecting vortices (in the ninth Section of the Second Book of the *Principia*) were mechanically erroneous; and in truth, it must be allowed that, besides several arbitrary assumptions, there are some errors of reasoning in them. But for the most part, the more enlightened Cartesians were content to accept Newton's account of the motions and forces of the solar system as part of their scheme; and to say only that the hypothesis of vortices explained the origin of the Newtonian forces; and that thus theirs was a philosophy of a higher kind. Thus it is asserted (Mém. Acad. 1734), that M. de Molières retains the beautiful theory of the Newton entire, only he renders it in a sort less Newtonian, by disentangling it from attraction, and transferring it from a vacuum into a plenum. This plenum, although not its native region, frees it from the need of attraction, which is all the better for it. These points were the main charms of the Cartesian doctrine in the eyes of its followers;—the getting rid of attractions, which were represented as revival of the Aristotelian "occult qualities," "substantial forms," or whatever else was the most disparaging way of describing the bad philosophy of the dark ages†;—and the providing some material intermedium, by means of which a body may affect another at a distance; and thus avoid the reproach urged against the Newtonians, that they made a body act where it was not. And we are the less called upon to deny that this last feature in the Newtonian theory was a difficulty, inasmuch as Newton himself was never unwilling to allow that gravity might be merely an effect produced by some ulterior cause.

With such admissions on the two sides, it is plain that the Newtonian and Cartesian systems would coincide, if the hypothesis of vortices could be modified in such a way as to produce the force of gravitation. All attempts to do this, however, failed: and even John Bernoulli, the most obstinate of the mathematical champions of the vortices, was obliged to give them up. In his Prize Essay for 1734, (on the Inclinations of the Planetary Orbits*,) he says (Art. VIII.), "The gravitation of the Planets towards the center of the Sun and the weight of bodies towards the center of the earth has not, for its cause, either the attraction of M. Newton, or the centrifugal force of the matter of the vortex according to M. Descartes;" and he then goes on to assert that these forces are produced by a perpetual torrent of matter tending to the center on all sides, and carrying all bodies with it. Such a hypothesis is very difficult to refute. It has been taken up in more modern times by Le Sage†, with some modifications; and may be made to account for the principal facts of the universal gravitation of matter. The great difficulty in the way of such a hypothesis is, the overwhelming thought of the whole universe filled with torrents of an invisible but material and tangible substance, rushing in every direction in infinitely prolonged straight lines and with immense velocity. Whence can such matter come, and whither can it go? Where can be its perpetual and infinitely distant fountain, and where the ocean into which it pours itself when its infinite course is ended? A revolving whirlpool is easily conceived and easily supplied; but the central torrent

---

† Acad. Sc. 1709. If we abandon the clear principles of mechanics, the writer says, "toute la lumière que nous pouvons avoir est eteinte, et nous voilà replongés de nouveau dans les anciennes ténèbres du Peripatetisme, dont le Ciel nous veuille preserver!"

It was also objected to the Newtonian system, that it did not account for the remarkable facts, that all the motions of the primary planets, all the motions of the satellites, and all the motions of rotation, including that of the sun, are in the same direction, and nearly in the same plane; facts which have been urged by Laplace as so strongly recommending the Nebular Hypothesis; and that hypothesis is, in truth, a hypothesis of vortices respecting the *origin* of the system of the world.

* *Novelle Physique Celeste,* Op. T. III. p. 163.

The deviation of the orbits of the planets from the plane of the sun's equator was of course a difficulty in the system which supposed that they were carried round by the vortices which the sun's rotation caused, or at least rendered evident. Bernoulli's explanation consists in supposing the planets to have a sort of *leeway (dérive des vaisseaux)* in the stream of the vortex.

† See *Phil. Ind. Sc.* B. III. c. ix. Art. 7.

of Bernoulli, the infinite streams of particles of Le Sage, are an explanation far more inconceivable than the thing explained.

But however the hypothesis of vortices, or some hypothesis substituted for it, was adjusted to explain the facts of attraction to a center, this was really nearly all that was meant by a vortex or a "tourbillon," when the system was applied. Thus in the case of the last act of homage to the Cartesian theory which the French Academy rendered in the distribution of its prizes, the designation of a Cartesian Essay in 1741 (along with three Newtonian ones) as worthy of a prize for an explanation of the Tides; the difference of high and low water was not explained, as Descartes has explained it, by the pressure, on the ocean, of the terrestrial vortex, forced into a strait where it passes under the Moon; but the waters were supposed to rise towards the Moon, the terrestrial vortex being disturbed and broken by the Moon, and therefore less effective in forcing them down. And in giving an account of a Tourmaline from Ceylon, (Acad. Sc. 1717) when it has been ascertained that it attracts and repels substances, the writer adds, as a matter of course, "It would seem that it has a vortex." As another example, the elasticity of a body was ascribed to vortices between its particles: and in general, as I have said, a vortex implied what we now imply by speaking of a central force.

4. In the same manner vortices were ascribed to the Magnet, in order to account for its attractions and repulsions. But we may note a circumstance which gave a special turn to the hypothesis of vortices as applied to this subject, and which may serve as a further illustration of the manner in which a transition may be made from one to the other of the two rival hypotheses.

If iron filings be brought near a magnet, in such a manner as to be at liberty to assume the position which its polar action assigns to them; (for instance, by strewing them upon a sheet of paper while the two poles of the magnet are close below the paper;) they will arrange themselves in certain curves, each proceeding from the N. to the S. pole of the magnet, like the meridians in a map of the globe. It is easily shewn, on the supposition of a magnetic attraction and repulsion, that these *magnetic curves,* as they are termed, are each a curve whose tangent at every point is the direction of a small line or particle, as determined by the attraction and repulsion of the two poles. But if we suppose a *magnetic vortex* constantly to flow out of one pole and into the other, in streams which follow such curves, it is evident that such a vortex, being supposed to exercise material pressure and impulse, would arrange the iron filings in corresponding streams, and would thus produce the phenomenon which I have described. And the hypothesis of *central torrents* of Bernoulli or Le Sage which I have referred to, would, in its application to magnets, really become this hypothesis of a magnetic vortex, if we further suppose that the matter of the torrents which proceed to one pole and from the other, mingles its streams, so as at each point to produce a stream in the resulting direction. Of course we shall have to suppose two sets of magnetic torrents;—a boreal torrent, proceeding to the north pole, and from the south pole of a magnet; and an austral torrent proceeding to the south from the north pole:—and with these suppositions, we make a transition from the hypothesis of attraction and repulsion, to the Cartesian hypothesis of vortices, or at least, torrents, which determine bodies to their magnetic positions by impulse.

Of course it is to be expected that, in this as in the other case, when we follow the hypothesis of impulse into detail, it will need to be loaded with so many subsidiary hypotheses, in order to accommodate it to the phenomena, that it will no longer seem tenable. But the plausibility of the hypothesis in its first application cannot be denied:—for, it may be observed, the two *opposite* streams would counteract each other so as to produce no local *motion*, only *direction*. And this case may put us on our guard against other suggestions of forces acting in curve lines, which may at first sight appear to be discerned in magnetic and electric phenomena. Probably such curve lines will all be found to be only resulting lines, arising from the direct action and combination of elementary attraction and repulsion.

5. There is another case in which it would not be difficult to devise a mode of transition from one to the other of two rival theories: namely, in the case of the emission theory and the undulation theory of Light. Indeed several steps of such a transition have already appeared in the history of optical speculation; and the conclusive objection to the emission theory of light, as to the Cartesian theory of vortices, is, that no amount of additional hypotheses will reconcile it to the phenomena. Its defenders had to go on adding one piece of machinery after another, as new classes of facts came into view, till it became more complex and unmechanical than the theory of epicycles and eccentrics at its worst period. Otherwise, as I have said, there was nothing to prevent the emission theory from migrating into the undulatory theory, and as the theory of vortices did into the theory of attraction. For the emissionists allow that rays may *interfere;* and that these interferences may be modified by alternate *fits* in the rays; now these fits are already a kind of *undulation.* Then again the phenomena of polarized light shew that the fits or undulations must have a *transverse* character: and there is no reason why emitted rays should not be subject to *fits* of *transverse* modification as well as to any other fits. In short, we may add to the emitted rays of the one theory, all the properties which belong to the undulations of the other, and thus account for all the phenomena on the emission theory; with this limitation only, that the emission will have no share in the explanation, and the undulation will have the whole. If, instead of conceiving the universe full of a *stationary* ether, we suppose it to be full of etherial particles moving in every direction; and if we suppose, in the one case and in the other, this ether to be susceptible of undulations proceeding from every luminous point; the results of the two hypotheses will be the same; and all we shall have to say is, that the supposition of the emissive motion of the particles is superfluous and useless.

6. This view of the manner in which rival theories pass into one another appears to be so unfamiliar to those who have only slightly attended to the history of science, that I have thought it might be worth while to illustrate it by a few examples.

It might be said, for instance, by such persons, "Either the planets are not moved by vortices, or they do not move by the law by which heavy bodies fall. It is impossible that both opinions can be true." But it appears, by what has been said above, that the Cartesians did hold both opinions to be true; and one with just as much reason as the other, on their assumptions. It might be said in the same manner, "Either it is false that the planets are made to describe their orbits by the above quasi-Cartesian theory of Bernoulli, or it is false that they obey the Newtonian theory of gravitation." But this would be said quite erroneously; for if the hypothesis of Bernoulli be true, it is so because it agrees in its result with the theory of Newton. It is not only possible that both opinions may be true, but is certain that if the first be so, the second is. It might be said again, "Either the planets describe their orbits by an inherent virtue, or according to the Newton theory." But this again would be erroneous, for the Newtonian doctrine decided nothing as to whether the force of gravitation was inherent or not. Cotes held that it was, though Newton strongly protested against being supposed to hold such an opinion. The word *inherent* is no part of the physical theory, and will be asserted or denied according to our metaphysical views of the essential attributes of matter and force.

Of course, the possibility of two rival hypotheses being true, one of which takes the explanation a step higher than the other, is not affected by the impossibility of two contradictory assertions of the *same order* of generality being both true. If there be a new-discovered comet, and if one astronomer asserts that it will return once in *every* 20 years, and another, that is will return once in every 30 years, both cannot be right. But if an astronomer says that though its interval was in the last instance 30 years, it will only be 20 years to the next return, in consequence of perturbation and resistance, he may be perfectly right.

And thus, when different and rival explanations of the same phenomena are held, till one of them, though long defended by ingenious men, is at last driven out of the field by the pressure of facts, the defeated hypothesis is transformed before it is extinguished. Before it has disappeared, it has been modified so as to have all palpable falsities squeezed out of it, and subsidiary provisions added, in order to reconcile it with the phenomena. It has, in short, been penetrated, infiltrated, and metamorphosed by the surrounding medium of truth, before the merely arbitrary and erroneous residuum has been finally ejected out of the body of permanent and certain knowledge.

TRINITY LODGE,   W. W.
    April 15, 1851.